Dedication

In dedication to Dr. Phil Rutledge,

Who never tired of showing us new ways to achieve environmental justice in our neighborhoods and communities—we are forever grateful for your conviction and the challenge you have left to us.

Fair and Healthy Land Use: Environmental Justice and Planning

CRAIG ANTHONY ARNOLD

TABLE OF CONTENTS

iv

The Intersection of Environmental Justice and Land Use Planning

Environmental justice is about the pursuit of fairness in environmental and land-use policies, especially fair treatment of all races, ethnic groups, and socioeconomic classes. Since 1980, new voices increasingly have been heard in opposition to the inequitable distribution of environmental harms and benefits by race and class in U.S. society. The environmental justice movement has used political activism, civil rights and constitutional law, environmental law, and new policies at all levels of government to seek fairness in environmental and land-use decisions. It has also sought to empower low-income communities of color to shape the environments in which they live and work.

Like many movements, the environmental justice movement does not have precise boundaries, nor does the term "environmental justice" have a precise definition. At its core, the concept of environmental justice is about the impacts of environmental and land-use policies on low-income communities of color. To a lesser degree, but still within the common meaning of the term, environmental justice addresses impacts on nonminority low-income communities and communities of color that are not composed of low-income residents. The term is sometimes used to address related communities, such as mixed-race or mixed-ethnicity communities and working class communities. This Planning Advisory Service Report focuses on both low-income communities and communities of color when the points are most relevant to both types of communities. In these circumstances, the points may also be relevant to mixed-race, mixed-ethnicity, and mixed-income neighborhoods. At other times, the report addresses specifically low-income communities of color that bear the greatest disproportionate impacts of land-use and environmental policies. Some variation may occur from locality to locality and region to region in the characteristics of the groups and neighborhoods most affected by environmental injustices.

Environmental justice and good land-use planning are inseparably connected. Land-use planning ideally should result from a fair and participatory process. A rich planning literature calls for robust public participation (even self-determination) in planning, the incorporation of social equity into plans, and vigilance in assessing the likely socioeconomic, racial, and ethnic impacts of land-use policies (American Planning Association 1994; Arnstein 1969; Beatley 1994, 87–101; Brooks 2002, 50–53, 107–18; Davidoff 1965; Forester, 1989; Krumholz and Forester 1990).

Plans and the zoning regulations that implement them ideally should segregate incompatible land uses, such as separating multifamily housing from industries that use and emit toxic substances. Good planning evaluates all of the likely significant impacts of various land uses on the human and natural environment, including cumulative impacts of concentrations of related land uses. Good plans provide for the social and physical needs of the people in the community, including public transportation, parks and open space, adequate facilities for public utilities, community centers, and the like. Good planning preserves and protects existing neighborhoods, with their heritages, social networks, and shared physical environments.

Likewise, poor land-use planning or the lack of land-use planning altogether has contributed to environmental injustice. Many of the environmental harms that low- and moderate-income minority communities face come from the proximate siting of locally unwanted land uses ("LULUs"), such as hazardous waste incinerators, solid waste landfills, and facilities that store, emit, or dispose of toxic substances. Moreover, studies of the distribution of zoning by race and class demonstrate that local governments zone communities where low-income people of color live for intensive land uses, particularly industrial land uses.

Many low-income and minority communities lack adequate public infrastructure, such as parks, open space, and upgraded water and sewer facilities. Poor transportation planning has exposed minority and poor neighborhoods to higher concentrations of air pollutants than other populations while failing to provide these neighborhoods with sufficient public transportation options. Minorities, poor people, and working class people complain that they have little say in shaping land-use plans and policies for their neighborhoods and that the process discourages their participation.

This PAS Report provides ideas and tools for integrating environmental justice into land-use planning. It is designed to assist planners and planning officials in thinking about environmental justice issues at all levels of decisions about land use, from the macro (e.g., the development of comprehensive plans for a city, county, or region) to the micro (e.g., the selection of mitigation conditions to be included in particular permits or other approvals for specific land-use projects).

This report attempts to provide a systematic---although certainly not exhaustive---way of thinking about the environmental justice issues in land-use decisions. It identifies the core principles of land-use planning and regulation that promote community participation, equitable treatment of all people, and a healthy, safe, and vibrant environment for community residents. It addresses comprehensive and neighborhood planning, smart growth and environmental justice, the use of environmental justice audits, zoning and advanced zoning techniques, land-use permits and approvals, exactions, public participation, environmental impact assessment, infrastructure, redevelopment, brownfields, and obstacles to incorporating environmental justice principles in land-use policies and decisions. Importantly, the report provides its readers with checklists to use in addressing environmental justice issues in land use and gives examples of approaches used by communities that have linked environmental justice and land-use planning.

CHAPTER 1

Environmental Justice: What Is it?

The environmental justice movement arose in the 1980s as a grassroots challenge to the fundamental fairness of environmental and land-use policies and decisions in the U.S. With origins in "the civil rights movement, the grassroots anti-toxics movement of the 1980s, organizing efforts of Native Americans and labor, and, to a lesser extent, the traditional environmental movement," the environmental justice movement is a national network of grassroots groups fighting the disproportionate impact of environmental decisions, land-use policies, and regulatory processes on low- and moderate-income people and people of color (Rechtschaffen and Gauna 2002, 3, citing Cole and Foster 2001).

The identification of a particular "environmental justice" movement is often traced to protests against the siting of a landfill accepting polychlorinated biphenyls (PCBs) in primarily African-American Warren County, North Carolina, in 1982 (Cole and Foster 2001, 19). Although there were many prior examples of low-income and minority people contesting the environmental and land-use burdens in their communities, the Warren County protests created a "framing event" that brought public attention to the relationships among: 1) environmental and land-use policies and practices; 2) civil rights; and 3) social justice.

The Warren County protests led to two landmark studies on the distribution of environmental harms by race and class. The first, by the U.S. General Accounting Office (GAO) (1983), investigated the distribution of the four major hazardous waste landfills in the Southeast. The GAO found that, of the four off-site hazardous waste landfills in the Environmental Protection Agency's (EPA's) eight-state Region IV, three were in communities in which African-Americans were a majority of the population. At the time, only about one-fifth of the population of Region IV was African-American. Subsequently, the United Church of Christ's Commission for Racial Justice (1987) undertook a significant, agenda-setting national study of demographic patterns associated with commercial hazardous waste facilities and uncontrolled toxic waste sites. The study found that three out of every five African-Americans and Hispanic-Americans nationwide were living in communities with uncontrolled toxic waste sites. Race was the most significant variable in the distribution of commercial hazardous waste facilities—more important than home ownership rates, income, and property values.

Since the 1980s, local conflicts over environmental justice have increased as the movement has grown. Some of these conflicts have captured national attention, heightening awareness of environmental justice as one of the important issues our society faces. Residents of Convent, Louisiana, a very-low-income, predominantly African-American community in the heart of the chemical-petroleum-industrial corridor between Baton Rouge and New Orleans, filed an environmental justice complaint with the U.S. Environmental Protection Agency against a proposal by Shintech Corporation for a polyvinyl chloride (PVC) manufacturing plant in Convent. Shintech withdrew its application after the EPA Administrator reversed state environmental approvals for failure to consider cumulative air pollution impacts (In *re Shintech Inc.* 1997; Mank 1999, 45-48). Latino residents of Kettleman City, California, many of whom are monolingual Spanish speakers and agricultural workers, challenged the adequacy of the English-only, highly technical environmental impact report accompanying the King's County Board of Supervisors' approval of a hazardous waste incinerator in Kettleman City. The proposed hazardous waste incinerator was dropped when a state court ruled for the community residents (*El Pueblo Para el Aire y Agua Limpio v. County of Kings* 1991; Cole and Foster 2001, 1–9). The primarily African-American community of Chester, Pennsylvania, in the Philadelphia metropolitan area, reacted to the high concentration of waste sites in this area by suing the state environmental agency under federal civil rights statutes for disparate impact in its approval of waste sites. Ultimately some of the cases brought by Chester environmental justice groups against waste facilities settled, the state agency denied environmental permits for a waste facility, and the U.S. Supreme Court declared the case against the agency to be moot (*Chester Residents Concerned for Quality Life v. Seif* 1997 and 1998; Cole and Foster 2001, 34–53). While the outcomes of other environmental justice conflicts have been similarly mixed, these high-profile examples illustrate the growing attention to fairness in environmental and land-use decisions.

Thus, the concern for environmental justice is a real, powerful, and pervasive force in U.S. society, even though many barriers exist to achieving environmental justice. One such barrier is the scope of the concept itself. The meaning of "environmental justice" and the goals of the environmental justice movement are complex and multifaceted. Evidence of environmental injustice depends on how the problem is defined and what methods are used to measure it. Causes of environmental injustice are hotly disputed. There are common principles or goals in the environmental justice movement, but specific goals are often context-specific, varying from community to community and from problem to problem, as would be expected from a grassroots-based movement. The types of injustices that arguably might be occurring include racism, discrimination against the poor, hostility to immigrants and/or people for whom English is not their primary language, exploitation of workers, the effects of historic segregation, lack of equal economic and political opportunity for all, and inattention to nonparticipants in the political system. Structural inequity and biases may exist inherently in our system of environmental law and regulation, our system of land-use planning and regulation, our political system, our educational system, our psychological predispositions, our social dynamics, and/or our economic system and its market dynamics. In other words, there is no single environmental justice problem or single source of environmental justice problems that creates an opportunity for a "quick fix." This complexity has impacts on land-use planning and decision making, as discussed below.

WHAT IS THE "ENVIRONMENT" IN ENVIRONMENTAL JUSTICE?

The concept of environmental justice extends to many different aspects of the environment. It encompasses exposure to pollution. However, pollution can include air pollution ranging from toxics in the air emitted by industry to particulate matter concentrated in inner cities by air basin dynamics, construction activity, and various aspects of urban life. Pollution can include water pollution ranging from heavy metals that accumulate in fish to fecal coliform in surface waters that receive sewage overflows and urban runoff. Pollution can include noise pollution from heavy truck traffic or manufacturing processes, litter, odor, toxic chemicals handled by workers, and spills of hazardous substances. An issue related to exposure to pollution is the disproportionate underenforcement of environmental laws in low-income and minority communities and the disproportionately lower penalties for environmental law violations that occur in low-income and minority communities.

Environmental justice also encompasses the siting of locally unwanted land uses (LULUs). Many of these LULUs, such as hazardous waste incinerators, hazardous waste storage and disposal facilities, solid waste landfills, sewage treatment facilities, power plants, and refineries, involve increased exposure to pollutants or increased risk of exposure to pollutants. Other LULUs, such as landfills, recycling facilities, and warehouses, may degrade the physical environment of a neighborhood and contribute to the decline of the community's sense of place. Still other LULUs, such as liquor stores, halfway houses, groups homes, and jails and prisons, may undermine a neighborhood's safety, stability, property values, and sense of community and place, if these LULUs are overconcentrated in the area.

More broadly, environmental justice encompasses land-use patterns generally, especially the location of industrial and commercial land uses with substantial adverse impacts in low-income and minority communities. However, the lack of certain land uses, facilities, and physical and social infrastructure is also an environmental justice issue. These often

A chrome plating facility located next to homes in Barrio Logan emitted very high levels of hexavalent chromium causing severe respiratory illness for neighborhood children. The Environmental Health Coalition of National City, California, organized community residents and worked with local and state officials to shut down and cleanup the facility.

Environmental Health Coalition

underprovided community resources include parks and open space, public transportation options, up-to-date utility services (including water supply distribution systems, sewer systems, and stormwater drainage systems), healthy streams and rivers, community centers and recreational facilities, and well-landscaped and well-maintained streets and sidewalks.

Some characterize the undersupply of affordable housing for low- and moderate-income people as an environmental justice issue. Some characterize the presence of contaminated or potentially contaminated, underproductive properties—brownfields— in low-income and minority communities as an environmental justice issue. Some characterize urban sprawl, with its siphoning of tax base, financial resources, and jobs away from cities and into suburbs and its segregating effect, as an environmental justice issue. Some see the lack of good-paying, safe, and healthy jobs in and near low-income and minority neighborhoods as an environmental justice issue. Finally, some in the environmental justice movement draw attention to low-income and minority communities' disproportionately lower degree of control of, and access to, natural resources.

The concepts of what is "just" with respect to environmental conditions are as diverse and broad as the scope of environmental conditions receiving attention.

WHAT IS THE "JUSTICE" IN ENVIRONMENTAL JUSTICE?

The concepts of what is "just" with respect to environmental conditions are as diverse and broad as the scope of environmental conditions receiving attention. One study of environmental justice stated there were seven different meanings of "fairness" that could apply merely to the issue of siting LULUs (Been 1993). Another study demonstrated that people can think of environmental justice issues as issues about the evidence of environmental justice, about political power, about legal rights, about enforcement of environmental laws, about economic resources and markets, or about planning and land-use controls (Arnold 1998). Yet another study of the "taxonomy" of environmental justice divided the concept into distributive justice, procedural justice, remedial (restorative) justice, and social justice (Kuehn 2000).

The broad scope of environmental justice as a concept is evident in the principles embraced by a national conference of grassroots activists and leaders of environmental justice groups in 1991. The First National People of Color Environmental Conference (1991) adopted a document containing 17 principles for environmental justice that would "serve as a defining document for the growing grassroots movement for environmental justice." These principles range from basing public policy on "mutual respect and justice for all peoples, free from any form of discrimination or bias," to rights of all peoples to environmental self-determination and participation at every level of decision making, to "the right to ethical, balanced and responsible uses of land and renewable resources in the interest of a sustainable planet for humans and other living things." They call for the clean-up and revitalization of urban and rural communities, damages for victims of environmental injustice, environmental and social justice education, and safe and healthy work environments. They also oppose the production of toxic, hazardous, or radioactive substances, the "destructive operations of multinational corporations," human rights abuses, and military occupation and repression. See the sidebar for the full list.

ARE THERE DISPARITIES? WHY DO THEY EXIST?

One of the complicating aspects of addressing environmental justice concerns is the debate over evidence. Is there adequate evidence that racial or class disparities in environmental conditions exist? Moreover, even if there is sufficient evidence of disparities, what is the cause (or what are the causes) of those disparities? Disagreements over evidence and causes characterize

PRINCIPLES OF ENVIRONMENTAL JUSTICE

The National People of Color Environmental Leadership Summit held on October 24-27, 1991, in Washington DC, drafted and adopted 17 principles of Environmental Justice. Since then, *The Principles* have served as a defining document for the growing grassroots movement for environmental justice.

PREAMBLE

WE, THE PEOPLE OF COLOR, gathered together at this multinational People of Color Environmental Leadership Summit, to begin to build a national and international movement of all peoples of color to fight the destruction and taking of our lands and communities, do hereby re-establish our spiritual interdependence to the sacredness of our Mother Earth; to respect and celebrate each of our cultures, languages and beliefs about the natural world and our roles in healing ourselves; to insure environmental justice; to promote economic alternatives which would contribute to the development of environmentally safe livelihoods; and, to secure our political, economic and cultural liberation that has been denied for over 500 years of colonization and oppression, resulting in the poisoning of our communities and land and the genocide of our peoples, do affirm and adopt these Principles of Environmental Justice:

1) **Environmental Justice** affirms the sacredness of Mother Earth, ecological unity and the interdependence of all species, and the right to be free from ecological destruction.

2) **Environmental Justice** demands that public policy be based on mutual respect and justice for all peoples, free from any form of discrimination or bias.

3) **Environmental Justice** mandates the right to ethical, balanced and responsible uses of land and renewable resources in the interest of a sustainable planet for humans and other living things.

4) **Environmental Justice** calls for universal protection from nuclear testing, extraction, production and disposal of toxic/hazardous wastes and poisons and nuclear testing that threaten the fundamental right to clean air, land, water, and food.

5) **Environmental Justice** affirms the fundamental right to political, economic, cultural and environmental self-determination of all peoples.

6) **Environmental Justice** demands the cessation of the production of all toxins, hazardous wastes, and radioactive materials, and that all past and current producers be held strictly accountable to the people for detoxification and the containment at the point of production.

Source: www.ejnet.org/ej/principles.html

7) **Environmental Justice** demands the right to participate as equal partners at every level of decision-making, including needs assessment, planning, implementation, enforcement and evaluation.

8) **Environmental Justice** affirms the right of all workers to a safe and healthy work environment without being forced to choose between an unsafe livelihood and unemployment. It also affirms the right of those who work at home to be free from environmental hazards.

9) **Environmental Justice** protects the right of victims of environmental injustice to receive full compensation and reparations for damages as well as quality health care.

10) **Environmental Justice** considers governmental acts of environmental injustice a violation of international law, the Universal Declaration On Human Rights, and the United Nations Convention on Genocide.

11) **Environmental Justice** must recognize a special legal and natural relationship of Native Peoples to the U.S. government through treaties, agreements, compacts, and covenants affirming sovereignty and self-determination.

12) **Environmental Justice** affirms the need for urban and rural ecological policies to clean up and rebuild our cities and rural areas in balance with nature, honoring the cultural integrity of all our communities, and providing fair access for all to the full range of resources.

13) **Environmental Justice** calls for the strict enforcement of principles of informed consent, and a halt to the testing of experimental reproductive and medical procedures and vaccinations on people of color.

14) **Environmental Justice** opposes the destructive operations of multi-national corporations.

15) **Environmental Justice** opposes military occupation, repression and exploitation of lands, peoples and cultures, and other life forms.

16) **Environmental Justice** calls for the education of present and future generations which emphasizes social and environmental issues, based on our experience and an appreciation of our diverse cultural perspectives.

17) **Environmental Justice** requires that we, as individuals, make personal and consumer choices to consume as little of Mother Earth's resources and to produce as little waste as possible; and make the conscious decision to challenge and reprioritize our lifestyles to insure the health of the natural world for present and future generations.

both general policy discussions and specific conflicts over individual projects or communities.

Many experts have conducted studies of environmental and land-use disparities, with varying results. (Many of these studies are listed in this report's list of references.) One problem, as might be expected, is that different studies used different methodologies and analytical tools. The geographic unit for environmental conditions or exposure to environmental harms differed by study, some using census tracts, others using ZIP codes, and others using more sophisticated socio-spatial analysis. Similarly, different studies evaluated different kinds of environmental problems or conditions, ranging from landfills, to sites storing or disposing of hazardous waste, to air pollution, and so forth. Some studies evaluated the distribution of then-existing land uses or environmental conditions, whereas others evaluated changes in land uses or environmental conditions over time.

A major environmental justice issue is the proximity of low-income children of color to sources of pollution.

The studies do not present a consistent pattern of correlation of specific community characteristics to disproportionately adverse environmental conditions. Thus, it is difficult to say whether a community's percentage of African-American residents, percentage of Hispanic residents, median income level, percentage of people living in poverty, voter participation rate, percentage of residents who do not speak English as their primary language, degree of ethnic stability or transition, other characteristics, or combination of characteristics is the best predictor of disproportionately adverse environmental conditions. In addition, there is sharp disagreement over whether LULUs are placed in low-income and minority neighborhoods (whether due to racism, class bias, cheap land, low levels of political opposition, or other factors), or whether low-income and minority people move to places where LULUs already exist (whether due to cheap housing, proximity to jobs, discrimination in housing markets elsewhere, or other factors).

While land-use decision makers should be aware of, and likely will encounter, some of the debates over evidence and causation, they should not be distracted by these debates for three reasons.

1. All of the studies, taken together, provide ample evidence that many low-income, high-minority communities face worse environmental and land-use conditions than other communities.

2. Environmental justice principles are good principles of planning and land-use practice, regardless of whether they attempt to remedy past problems or not. Environmental justice principles assert that no person or neighborhood should be burdened by harmful environmental conditions and that all persons should be treated fairly and should have the opportunity for full, meaningful participation in the decisions affecting the health, safety, and identity of their community. As will be discussed subsequently in Chapter 2, these concepts are at the core of the purposes that land-use planning and regulation serve in our society.

3. Discussion about what happened in the past does not necessarily address the question of what should happen now or in the future. Planning and land-use decision making are about what should happen now or in the future. Indeed, it may matter very little which type of land use came first because land-use patterns change, property owners seek to take advantage of new opportunities by changing how they use their land, and public officials and community residents periodically engage in developing new plans for the future. The frequent changes in the characteristics of a locality's land uses are common phenomena. A wharf once used for shipping and warehousing, for example, now becomes a center of shops, hotels, restaurants, and tourist attractions. Warehouses are converted to condominiums. Agricultural land gives way to residential development. A property owner seeks to convert a vacant florist's shop to an automotive repair facility. Another property owner seeks to convert a home into a set of professional offices. Simply because low-income and minority housing, on one hand, and industries and LULUs, on the other hand, are located in close proximity to (or even interspersed among) one another, there is no reason to expect that these existing land-use patterns will continue in perpetuity. As conditions change and new land-use opportunities arise, environmental justice principles in plans and standards for making land-use decisions can guide the direction in which low-income and minority areas change without the necessity of resolving all disputes about past practices.

Environmental justice principles are good principles of planning and land-use practice, regardless of whether they attempt to remedy past problems or not.

WHAT IMPACT DOES ENVIRONMENTAL JUSTICE HAVE ON PUBLIC POLICY AND LAW?
Despite the many possible meanings of environmental justice (and injustice), the concept is increasingly playing an important role in public policy and law. Environmental justice advocates are seeking policy changes and legal rights and remedies in many different areas, and at many different levels of decision making that affect environmental conditions in low-income communities and communities of color.

Lawsuits challenging the disproportionate effects of government decisions on low-income and minority communities have increased in frequency, at least if studies of reported court decisions are any indication (Binder 1995; Binder 2000; Binder 2005). Litigation under the Equal Protection Clause of the U.S. Constitution or under federal civil rights statutes, however, has not fared well for environmental justice advocates because of the requirement that plaintiffs prove discriminatory intent by government officials, not merely the discriminatory, or disparate, impact of the outcome. Litigation under federal and state environmental statutes, though, has proven to be more successful for environmental justice groups. Plaintiffs may be successful in proving that government decision makers did not adequately consider the environmental impacts of their decisions on low-income and minority

Increasingly, states are developing their own environmental justice policies, some with far-reaching implications for land-use planning and regulation.

communities, did not follow mandatory procedures, or did not properly apply standards aimed at protecting public health and safety. Nonetheless, for many environmental justice groups, litigation may be merely one of several strategies designed to put pressure on decision makers, create adverse publicity for project proponents and supporters, increase the costs of seeking to place unwanted land uses in low-income and minority communities, and empower local residents (Cole 1992). Interestingly, some of the lawsuits brought to stop waste facilities and other unwanted land uses in low-income, minority communities have involved challenges to local zoning or land-use permit approvals.

A major development in environmental justice policy occurred in 1994, when President Clinton signed Executive Order 12898, directing all federal agencies to address environmental justice in federal agency actions and to develop strategies for doing so. (The full text can be found in Appendix A of this PAS Report.) The Order mandates that:

> each Federal agency shall make achieving environmental justice part of its mission by identifying and addressing, as appropriate, disproportionately high and adverse human health or environmental effects of its programs, policies, and activities on minority populations and low-income populations in the United States." (Executive Order 12898, February 11, 1994, Section 1–101)

A 2001 study of federal agency responses to the Executive Order shows the impact has been mixed (Binder et al. 2001). On one hand, the researchers did not find any examples of a project being denied or a program being dropped for environmental justice reasons. In addition, most of the agency's environmental justice initiatives involved either repackaging existing programs or undertaking discrete new projects. Comprehensive restructuring of agency programs and development of regulatory systems to achieve environmental justice has either not occurred at all or occurred only in rare, limited instances. On the other hand, though, the researchers found that many agencies substantially increased or improved public participation in gathering and disseminating information and in reaching out to low-income and minority communities. They also found that some agencies invested substantial resources in particular environmental justice issues, such as brownfield redevelopment and lead-paint remediation programs. They also noted that agency decision makers likely are more aware of environmental justice issues when implementing existing programs and regulations.

Increasingly, states are developing their own environmental justice policies, some with far-reaching implications for land-use planning and regulation (Bonorris et al. 2004; Rechtschaffen 2003; National Academy of Public Administration 2002; Rechtschaffen and Gauna 2002, 414–16). California, for example, adopted legislation requiring that the state's general plan guidelines include an environmental justice section encouraging localities to adopt plans preventing overconcentration of industrial facilities near residences and schools and inequitable distribution of community-enhancing public facilities and services (California Governor's Office of Planning and Research 2003, 20-31). Some states have enhanced the public participation components of state and local permitting to increase public awareness of, and involvement in, environmental and land-use permits. Some states require studies of the distribution of environmental hazards by race, income, or other social factors, or the consideration of the impacts in burdened communities of proposed state actions or permits. Others have laws or regulations to address or prevent overconcentration of certain land uses. (See Figure 1-1, which shows just such overconcentration

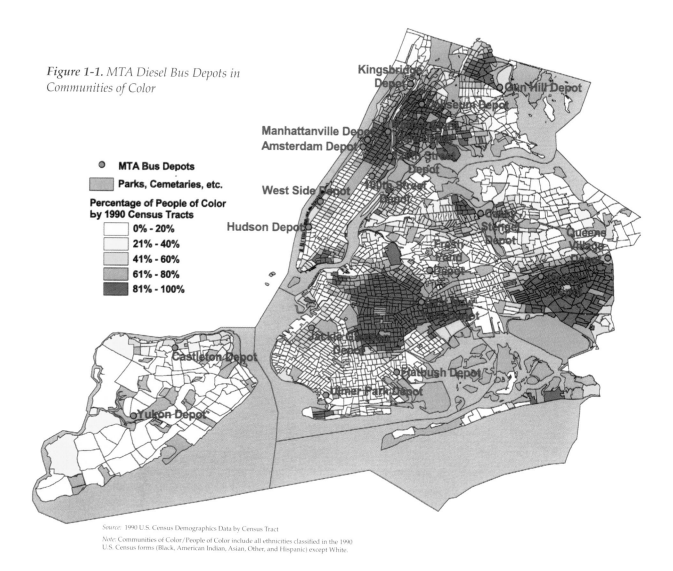

Figure 1-1. MTA Diesel Bus Depots in Communities of Color

Source: 1990 U.S. Census Demographics Data by Census Tract

Note: Communities of Color/People of Color include all ethnicities classified in the 1990 U.S. Census forms (Black, American Indian, Asian, Other, and Hispanic) except White.

of Metropolitan Transit Authority bus depots in communities of color.) Some have incorporated environmental justice into planning for certain facilities, such as transportation, hazardous waste sites, or power plants. Some states have created community advisory boards or task forces, environmental justice offices, or environmental justice centers to address environmental justice concerns.

These changes are having impacts. In 2005, a Rhode Island court found that the state environmental agency violated Rhode Island's Industrial Property Remediation and Reuse Act when if failed to consider issues of environmental equity in remediating and approving a former landfill for a public school site despite evidence of soil toxics, as well as failing to provide adequate public notice, hearings, and access to records (*Hartford Park Tenants Association v. Rhode Island Department of Environmental Management* 2005).

Most importantly, though, grassroots environmental justice groups are expressing their concerns and pursuing their goals at all levels of government and in a wide array of decisions that affect environmental conditions in their communities. Throughout the U.S., environmental justice advocates pursue political activism, opposition to specific facilities, and litigation. Many grassroots groups are also seeking a voice in local land-use planning, policy making, and decision making (Arnold 1998, 98–105). They are assessing the conditions in their own neighborhoods and engaging in community-based planning exercises to offer their own vision of what their neighborhoods could be, free from environmental injustice and land-use inequity. They are advocating for enhanced facilities and services in their communities. From New York City to Chicago to Austin to Denver to San Diego, low-income and minority community residents are influencing local land-use politics and policies.

CHAPTER 2

Environmental Justice
and Land Use

Planning for environmental justice is critical to a society that seeks to achieve environmental justice. In fact, the very nature of planning itself calls for the adoption and implementation of environmental justice principles and policies. Planning is the process of identifying goals for the future, developing policies or plans for achieving the goals, and fashioning specific mechanisms for implementing the plans. It also contains phases of pre-plan study and post-plan monitoring and feedback (So and Getzels 1988, 10–11). The American Planning Association (1979) has defined planning as "a comprehensive, coordinated and continuing process, the purpose of which is to help public and private decision makers arrive at decisions which promote the common good of society." Some of the public interest goals served by planning are health, safety, convenience, efficiency, natural resource conservation, environmental quality, social equity, social choice, amenity, and morals (Chapin and Kaiser 1979, 48).

THE PROMISE AND FAILURE OF PLANNING

Urban planning, throughout much of its U.S. history, has offered the promise of a high-quality physical and social environment for the least advantaged in society. Although at times concerned with long-term utopian visions of a well-planned society, planners often focused on specific practical problems that dominated the public agenda of the times:

- health and safety issues like public sanitation, tenement housing conditions, and sewage in the latter half of the nineteenth century;

- aesthetically pleasing infrastructure that promoted civic life and equality (e.g., parks, civic centers, streets, and transportation), such as the goals of the City Beautiful movement at the turn of the century;

- the economic and social problems presented by uncoordinated development and inadequate municipal services in the face of urbanization in the early twentieth century; and

- the problems of poverty, slums, and housing shortages and conditions from the 1930s through the 1960s (Young 1996, Sections 1.04–1.07, 10–13; So and Getzels 1988, 26–28, 30–46, 61–67).

Our modern land-use planning and regulatory system, including zoning, arose in response, in part, to substandard and unhealthy conditions of housing among the poor and racial and ethnic minorities.

© iStockphoto.com/Perry Knoll

Thus, the theory and practice of planning land uses to promote the common good have been marked by elements of environmental justice, even though the terminology and express identification of environmental justice principles are more recent.

The Legal Justification for Planning

A primary justification for land-use planning and regulation has been that land-use plans and controls protect people from living near or among environmentally harmful land uses. Land-use regulation, from its early history, prevented incompatible, noxious uses from interfering with the private enjoyment of property, private property values, and public health and safety (Mandelker 2003, Section 2.04, 2–6 to 2–8).

In 1926, the Supreme Court upheld the constitutionality of an early zoning ordinance, despite the Court's strong antiregulatory, pro-private property jurisprudence of that era. In the landmark case of *Village of Euclid v. Ambler Realty Co.* (272 U.S. 365 (1926)), the Court made an analogy between: (1) regulatory prohibitions of nonresidential uses in residential neighborhoods and of building structures that did not conform to height limits, construction standards, and setbacks; and 2) common law restrictions on nuisances. In discussing the prohibition of industrial uses in residential neighborhoods, the Court stated:

> Thus the question whether the power exists to forbid the erection of a build-ing of a particular kind or for a particular use, like the question whether a particular thing is a nuisance, is to be determined, not by an abstract consid-eration of the building or of the thing considered apart, but by considering it in connection with the circumstances and the locality. . . . A nuisance may be merely a right thing in the wrong place, like a pig in the parlor instead of the barnyard. (388)

Nearly 50 years later, a very different Supreme Court, deciding the validity of a zoning ordinance restricting certain areas to single-family residences, communicated its approval of planning and zoning to promote residential enclaves "where family values, youth values, and the blessings of quiet se-clusion and clean air make the area a sanctuary for people" (*Village of Belle Terre v. Boraas*, 416 U.S. 1, 9 (1974)). These two decisions from the nation's highest court, arising under different historical conditions and jurisprudential eras, reflect the widespread faith, or at least hope, in the role and efficacy of land-use planning and regulation to provide people with a high-quality living environment. At the very least, land-use planning offers the promise of healthy, safe, vibrant communities for all people.

The History of Unjust Land-Use Practices

The practice of land-use planning and regulation has all too often been characterized by environmental injustice. Despite the use of planning and zoning to protect many residential neighborhoods from incompatible uses and to provide them with environmental amenities and public infrastructure, the system has failed many low-income and minority neighborhoods. There has been an appalling tendency either to treat low-income and minority neighborhoods as "barnyards" suitable for placing the "pigs" of intensive or unwanted land uses or to treat people of color and poor people as "pigs" who should be placed in the "barnyards" of industrial and commercial areas or areas without adequate schools, parks, and the like. In fact, the Supreme Court's opinion in *Euclid*, which used the pig-parlor-barnyard analogy, spe-cifically identified apartment buildings as "parasites" that should be kept out of single-family residential neighborhoods (394–95). The implication was that multifamily housing is more like an industrial land use than it is like single-family housing.

The racial, ethnic, and class injustices of land-use planning and regulation have had many manifestations. In the early twentieth century, cities—mostly in the South but as far north as Indiana and Maryland and as far west as Texas and Oklahoma—adopted ordinances segregating African-Americans and whites by geographic areas of the city (Ellickson and Been 2005, 692). Although the Supreme Court struck down the Louisville, Kentucky, resi-dential segregation ordinance in 1917 (*Buchanan v. Warley*), explicitly racial zoning ordinances persisted into the late 1940s (Ellickson and Been 2005, 693). Race-specific zoning was combined in some cases with industrial zon-ing policies that directly placed industries and minority residences in the same areas. For example, Austin, Texas, planned the area of East Austin in 1928 as a "negro district" that would host most of Austin's industrial uses next to housing for African-Americans, and the city's first zoning map in 1931 reflected this plan (Greenberger 1997).

A more widespread and persistent zoning practice than race-specific districting has been the use of exclusionary zoning techniques (Dubin 1993; Collin 1992, 507–09; Ellickson and Been 2005, 691). These techniques

City of North Las Vegas

Suburban sprawl is an environmental justice when it shifts jobs, retail shopping, infrastructure, and tax revenues from central cities to more exclusive, less diverse suburban communities.

Sprawl allows for "white flight" from urban areas and demands significant investment of a region's public resources in suburban areas.

indirectly exclude certain groups from particular communities or areas by controlling the type of housing development that occurs in those areas. These techniques include large-lot zoning, low-density zoning, growth moratoria or tempo controls that limit the supply of new housing, costly exactions and development conditions, and lack of lots zoned for multifamily housing (Callies et al. 1994, 431–34; Selmi and Kushner 2004, 519–22; Ellickson and Been 2005, 691). Exclusionary zoning has the effect of limiting or precluding affordable housing in a community. Thus, it keeps out those who cannot afford higher-cost housing, including low- and moderate-income people, racial and ethnic minorities, young and elderly couples, single persons, and large families (*Southern Burlington County N.A.A.C.P. v. Township of Mount Laurel* 1975).

Land-use policies encouraging or facilitating suburban sprawl have contributed to racial and income disparities in U.S. society (Pulido 2000; Bullard 2000; Kushner 2002–2003; Hutch 2002). Sprawl allows for "white flight" from urban areas and demands significant investment of a region's public resources in suburban areas. Many suburban communities have their own separately incorporated municipalities with independent political, land-use, tax, and fiscal powers. The result can be municipal financial stress in central cities and underinvestment in inner-city facilities and services.

Even aside from interjurisdictional competition for resources, many cities have not provided municipal services and facilities to low-income and minority areas at the same level or to the same degree as they have to other parts of their cities (Haar and Fessler 1986; Bond 1976; Garcia and Flores 2005). In addition, public investments in inner-city areas may harm low-income and minority communities when they take the form of redevelopment projects that displace community residents or even destroy entire communities (Jacobs 1961, 137). Revitalization of inner-city areas, in some cases, has amounted to gentrification, raising housing and other costs of living in the area, reducing the supply of affordable housing, and forcing out low-income and minority people (Kelly and Becker 2000, 347–48; McFarlane 2006).

In addition to the segregating effects of zoning and land-use practices, these practices have also had the effect of burdening low-income and minority communities with unwanted land uses and environmental harms. Yale Rabin (1990) has documented the rezoning of low-income and minority areas to accommodate industrial and similarly intensive land uses. He calls this

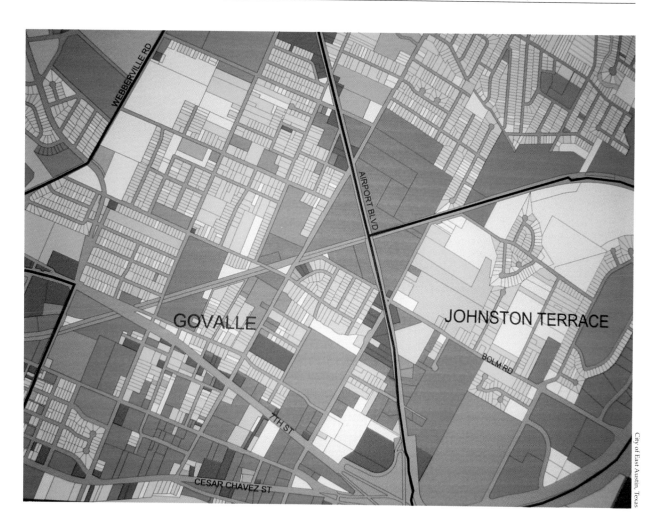

practice "expulsive zoning" because of its effect of driving out residents and land uses that can afford to move elsewhere. The result is that poor people and minorities who lack housing opportunities elsewhere are located near health-harming, community-degrading land uses. The zoning or rezoning of low-income and minority communities to allow industrial land uses has contributed to decisions to locate waste facilities and other locally unwanted land uses (LULUs) in these communities (e.g., *R.I.S.E., Inc. v. Kay* 1991; *Lake Lucerne Civic Association v. Dolphin Stadium Corporation* 1992; National Academy of Public Administration 2003).

Finally, disparities in participation by the poor and minorities in land-use planning and regulatory processes are common. Studies reflect that land-use decision makers tend to be white males, middle-aged or older, at least in higher proportion than their share of the local population, and many of them tend to be real estate professionals with a vested interest in land-use decisions (Anderson and Sass 2004; Anderson and Luebbering 2006; Sanders and Getzels 1987; National Academy of Public Administration 2003, 50–51). Low-income people of color not only have not held positions on planning and zoning commissions, city councils, and similar boards in any significant numbers, but they also have participated in public hearings and planning forums in relatively low numbers. The timing and location of meetings, limited access to information, language and education barriers, and perceptions of powerlessness contribute to these low levels of participation.

The cumulative effect of these land-use policies—as well as private discrimination, racially restrictive covenants, redlining, housing development policy, inaccessible markets, and other practices—has been to create

An Austin, Texas, zoning map depicting the interspersing of industrial zoning (purple) and residential zoning (yellow) in the predominantly low-income and minority community of East Austin. The City of Austin, in cooperation with East Austin residents and the environmental justice group PODER (People Organized for Defense of Earth and her Resources), engaged in a comprehensive replanning and rezoning of the area to eliminate incompatible land use designations. See the sidebar in Chapter 4.

and sustain patterns of racial residential segregation in the U.S. over time (Massey and Denton 1993; Kushner 1979). A recent Brookings Institution report reveals concentrations of extreme poverty in America's inner cities, often predominated by racial and ethnic minorities (Berube and Katz 2005). Moreover, despite the partial integration of formerly all-white neighborhoods, very little desegregation of predominantly African-American areas has occurred (Ellickson and Been 2005, 697), and instead segregated middle-class black suburbs have emerged (Cashin 2004).

EMPIRICAL EVIDENCE OF DISPARITIES IN LAND-USE PATTERNS

Expulsive Zoning

Land-use patterns themselves make a compelling case for the equitable failure of land-use planning historically, as well as the need—and opportunity—to incorporate environmental justice principles into land-use planning and decisions. In 12 case studies, planning expert Yale Rabin (1990) documented that various cities nationwide rezoned neighborhoods of color to allow incompatible and noxious land uses, thereby displacing some residents and replacing them with new industrial and commercial activities that threatened the health, safety, quality, and character of the neighborhood for those residents least able to leave or find housing elsewhere. Rabin called this type of zoning "expulsive zoning." Rabin's study, however, did not attempt to quantify the distribution of zoning patterns in low-income neighborhoods of color and compare those distributions with zoning patterns of high-income white neighborhoods in the same cities.

Comparing Low-Income, High-Minority Census Tracts with High-Income, Low-Minority Census Tracts

A 1998 study of zoning patterns in 31 census tracts in seven cities did examine what Rabin did not—the percentage of low-income, high-minority census tracts zoned for industrial and other intensive land uses compared with that percentage in high-income, low-minority census tracts (Arnold 1998). The results showed a great disparity.

This study, which I conducted, measured the percentages of area in census tracts that local zoning ordinances had designated for each type of land-use It contained data from 31 census tracts in seven cities: Anaheim, California; Costa Mesa, California; Orange, California; Pittsburgh, Pennsylvania; San Antonio, Texas; Santa Ana, California; and Wichita, Kansas. I chose census tracts by reviewing 1990 U.S. Census Bureau data documenting racial composition, median household income, and percentage of persons below the poverty level for all the census tracts of each city. I specifically studied census tracts with populations either significantly above or significantly below the racial and class composition of the city. There were 19 low-income, high-minority census tracts, all of which had more than 150 percent of their respective city's percentages of people below poverty and people of color, except for two tracts in San Antonio and three tracts in Santa Ana. These five exceptions had less than 150 percent of the respective city's percentages of people of color due to the high number of people of color in those cities. Each of the five tracts had more than 85 percent people of color, and three of the tracts had 92 percent or more. In absolute measures, all low-income, high-minority tracts in all cities had more than 45 percent people of color, and 16 out of the 19 tracts had more than 69 percent. All low-income, high-minority tracts had more than 15 percent of their populations living below poverty level, and 13 out of the 19 tracts had 33 percent or more of their populations living below poverty level. I also selected 12 high-income, low-minority census tracts, all of which had less than 51 percent of the respective city's percentages for people living below poverty level and people of color. In absolute measures, all high-income, low-minority tracts in all cities had less than 27 percent people of color, and eight out of the 12 tracts had 14 percent or less. All high-income, low-minority tracts had less than 8 percent people living below poverty level, and 75 percent of these tracts had 4.5 percent or less that fit that criterion.

The data (summarized in Table 2-1 and Figures 2-1 through 2-4) show that low-income, high-minority neighborhoods in the cities studied are subject to more intensive zoning, on the whole, than high-income, low-minority neighborhoods. This conclusion is supported by data from across the various types of cities studied, regardless of the cities' geographic features, spatial development, population, political characteristics, and the like.

TABLE 2-1. EMPIRICAL STUDY OF LAND-USE PATTERNS: ZONING CLASSIFICATIONS
BY JURISDICTION AND CENSUS TRACT

LEGEND FOR TABLES AND GRAPHS

Symbol	Term
*	High-income, low minority census tract
#	Low-income, high-minority census tract
SFR	Single-family residential (includes low-density residential)
MFR	Multi-family residential (includes two-family residential, duplex residential, manufactured housing, mobile home residential, and medium- and high-density residential)
C	Commercial (includes business and professional)
I	Industrial
PD	Planned Development
O	Other (includes open space, park/recreation, country club, public use, government center and special [Pittsburgh])

ANAHEIM, CALIFORNIA, PERCENT OF CENSUS TRACTS BY AGGREGATED ZONING DESIGNATIONS

Tract	SFR	MFR	C	I	PD	O
219.04*	94.98	4.84	0.17	0	0	0
874.02#	22.74	25.42	16.99	23.74	11.12	0
874.03#	57.94	12.50	22.59	3.34	3.63	0

COST MESA, CALIFORNIA, PERCENT OF CENSUS TRACTS BY AGGREGATED ZONING DESIGNATIONS

Tract	SFR	MFR	C	I	PD	O
638.02*	57.82	5.05	16.67	0	0	20.46
637#	32.25	25.51	28.68	4.79	0	8.78

ORANGE, CALIFORNIA, PERCENT OF CENSUS TRACTS BY AGGREGATED ZONING DESIGNATIONS

Tract	SFR	MFR	C	I	PD	O
219.12*	25.89	0	0	2.84	49.83	21.44
762.04#	0	8.08	20.46	68.84	0	2.61

PITTSBURGH, PENNSYLVANIA, PERCENT OF CENSUS TRACTS BY AGGREGATED ZONING DESIGNATIONS

Tract	SFR	MFR	C	I	PD	O
1401.98*	42.57	7.02	0	0	2.96	47.44
1404*	66.02	23.41	0.73	0	0	9.84
1106*	6.82	22.28	0	0	0	70.90
509#	0	57.74	0	1.94	0	40.33
510#	0	4.63	0	0	57.19	38.19
1016#	0	31.74	0	0	56.71	1.58
2609.98#	50.64	1.70	1.35	1.21	0	45.10
2808#	5.94	13.88	0.74	50.11	12.28	17.05

TABLE 2-1. EMPIRICAL STUDY OF LAND-USE PATTERNS: ZONING CLASSIFICATIONS
BY JURISDICTION AND CENSUS TRACT *(continued)*

SAN ANTONIO, TEXAS, PERCENT OF CENSUS TRACTS BY AGGREGATED ZONING DESIGNATIONS

Tract	SFR	MFR	C	I	PD	O
1204*	Approximately 99.00	Approximately 0	1.00	0	0	0
1914.02*	95.22	1.98	2.81	0	0	0
1915.02*	89.92	6.07	4.00	0	0	0
1105#	9.79	34.92	6.43	48.30	0	0.56
13.05#	38.39	48.22	11.72	1.64	0	0.04
1307.85#	14.52	15.72	33.17	36.59	0	0
1702#	69.70	5.67	24.50	0	0	0.14

SANTA ANA, CALIFORNIA, PERCENT OF CENSUS TRACTS BY AGGREGATED ZONING DESIGNATIONS

Tract	SFR	MFR	C	I	PD	O
753.03*	81.05	1.59	16.67	0	00.69	
744.03#	3.43	2.82	0.65	90.54	2.56	0
749.01#	17.88	33.64	16.77	0	18.45	13.43
750.02#	0	12.34	48.30	0	13.20	26.07

WICHITA, KANSAS, PERCENT OF CENSUS TRACTS BY AGGREGATED ZONING DESIGNATIONS

Tract	SFR	MFR	C	I	PD	O
73.01*	67.95	5.59	9.77	0	0	16.68
74#	100.00	0	0	0	0	0
8#	0	94.36	5.65	0	0	0
41#	0	6.77	70.68	22.55	0	0
78#	68.03	19.59	5.85	6.52	0	0

Disparities in Industrial Zoning

With respect to industrial zoning, the most intensive land-use, 13 out of 19 low-income, high-minority census tracts had at least some industrial zoning, and in seven of those census tracts, the city had zoned more than 20 percent of the tract for industrial uses. In contrast, only one of the 12 high-income, low-minority census tracts contained any industrial zoning at all, and then only 2.84 percent of the tract was zoned industrial.

More specifically, 90.54 percent of Santa Ana tract #744.03 was zoned for industrial use. This census tract was home in 1990 to 4,862 people, of whom 74.9 percent are Hispanic. Nearly 70 percent of Orange tract #762.04, about 50 percent of both Pittsburgh tract #2808 and San Antonio tract #1105, and 36.59 percent of San Antonio tract #1307.85 were zoned for industrial use. These census tracts provided homes to between 2,700 people and 3,500 people each. Moreover, although the study did not include a quantified spatial distribution analysis of the industrial uses in comparison to the residential uses, a visual survey of the zoning maps revealed that industrial-use designations were close to residential-use designations, *often either across the street or in the same block.* Industrial zoning was interspersed with residential zoning in many of the tracts studied.

The zoning of low-income neighborhoods of color for industrial uses placed highly intensive activities near local residents' homes, creating the very sort of incompatibility of uses zoning is designed to prevent. (*Euclid,* 272 U.S. at 386) For example, among the "as of right" permitted uses in Pittsburgh tract #2808 were ammonia and chlorine manufacturing, automobile wrecking, blast furnace or coke oven, chemical manufacturing, iron and steel

TABLE 2-1. EMPIRICAL STUDY OF LAND-USE PATTERNS: ZONING CLASSIFICATIONS
BY JURISDICTION AND CENSUS TRACT *(continued)*

INDUSTRIAL ZONING BY CENSUS TRACTS

City	Census Tract	Percent of Persons of Color	Percent of Low-Income Persons	Percent of Tract Zoned for Industrial Use
Anaheim	219.04	Low	Low	0
	874.02	High	High	23.74
	874.03	High	High	3.34
Costa Mesa	638.02	Low	Low	0
	637	Medium	High	4.79
Orange	219.12	Low	Low	2.84
	762.04	High	High	68.84
Pittsburgh	1401.98	Low	Medium[1]	0
	1404	Low	Low	0
	1106	Low	Low	0
	509	High	High	1.94
	510	High	High	0
	1016	High	High	0
	2609.98	High	High	1.21
	2808	High	High	50.11
San Antonio	1204	Low	Low	0
	1914.02	Low	Low	0
	1915.02	Low to Medium	Low	0
	1105	High	High	48.30
	1305	High	High	1.64
	1307.85	High	High	36.59
	1702	High	High	0
Santa Ana	753.03	Low	Medium	0
	744.03	High	High	90.54
	749.01	High	High	0
	750.02	High	High	0
Wichita	73.01	Low	Low	0
	74	Low	Low	0
	8	High	High	0
	41	High	High	22.55
	78	High	High	6.52

manufacturing and processing, airplane factory or hangar, brewery, poultry slaughter, and machine shop, and among the conditional uses are atomic reactors, garbage and dead animal reduction, rubbish incineration, radio and television transmission and receiving towers, and storage of explosives and inflammables. San Antonio allowed acetylene gas manufacturing and storage, arsenals, blast furnaces, boiler works, cement or paving material mixing plants, creameries with on-premises livestock, forge plants, metal foundries, paper and pulp manufacturing, rock crushers, junk storage, tar-roofing manufacturing, and yeast plants, among others, in two of the census tracts studied. Even though nearly two-thirds of Orange census tract #762.04 was zoned for industrial manufacturing (M2), the city required many of the most intensive uses to obtain conditional use permits, thus at least theoretically allowing some level of monitoring and control of the impacts. Nevertheless, some of the conditionally permitted uses in Orange's M2 district were hazardous waste facilities, refuse transfer stations, blast furnaces and coke ovens, mineral extraction and production, and various

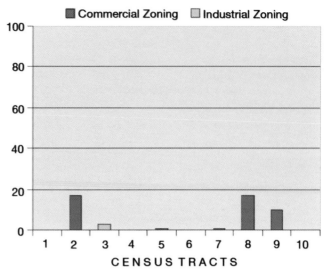

Figure 2-1. Percentages of High-Income, Low-Minority Census Tracts Zoned for Industrial and Commercial Land Uses

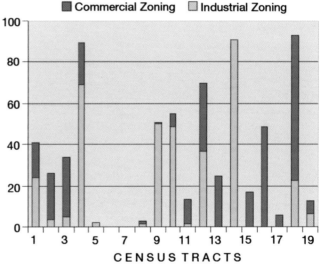

Figure 2-2. Percentages of Low-Income, High-Minority Census Tracts Zoned for Industrial and Commercial Land Uses

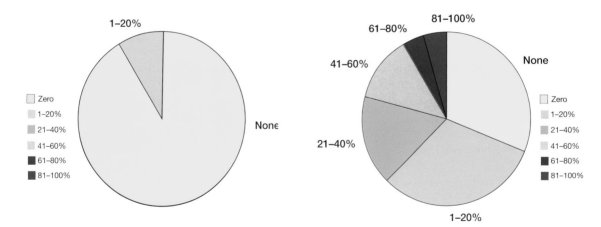

Figure 2-3. Percentages of High-Income, Low-Minority Census Tracts by Percentage of Industrial Zoning

Figure 2-3. Percentages of Low-Income, High-Minority Census Tracts by Percentage of Industrial Zoning

types of chemical production. Santa Ana zoned nearly 90 percent of census tract #744.03, containing nearly 5,000 residents, for light industrial activity. Although Santa Ana's light industrial zoning designation excluded hazardous and solid waste facilities and some hazardous industrial activities (e.g., acid manufacturing, gas and acetylene manufacturing, and metal smelters), it did not exclude large-scale industrial facilities that can overwhelm nearby residential uses, the use of toxic substances in light industrial activities, unsightly storage facilities and warehouses, or a high concentration of waste-producing facilities (e.g., automotive repair and service sites).

Commercial Zoning with High Impact

Commercial uses were also located in greater concentrations in low-income, high-minority neighborhoods than in high-income, low-minority neighborhoods. In 10 out of the 19 low-income, high-minority census tracts, at least 10 percent of the area was zoned for commercial use, and in seven of those tracts, at least 20 percent of the area was zoned for commercial use. In contrast, only two of the 12 high-income, low-minority census tracts had at least 10 percent of the area zoned for commercial use, and none had more than 20 percent commercial zoning.

Although the term "commercial" conjures up images of office buildings and retail stores that may create parking and scale/shadow impacts on neighboring residences but generally do not pose health hazards, the cities studied allowed in their various commercial districts uses far more intensive than offices and stores. For example, nearly 50 percent of Wichita tract #41 was zoned Central Business District, in which limited and general manufacturing, vehicle storage yards, warehousing, welding and machine shops, and vehicle repair uses were allowed by right, and solid waste incinerators, mining and quarrying, rock crushing, and oil and gas drilling were conditional uses. In about 30 percent of San Antonio tract #1307.85, permitted uses included electroplating, breweries, chicken hatcheries, poultry slaughter and storage, machine shop, and certain kinds of manufacturing, such as ice cream, ice, brooms, mattresses, paper boxes, candy, cigars, and refrigeration. Santa Ana's General Commercial (C2) districts could contain automotive garages, blueprinting and photo-engraving businesses, metal shops, automotive equipment wholesalers, research laboratories, farm products wholesalers, and tire recapping businesses, and the Central Business (C3) District could contain all of these land uses except automotive garages. These "commercial" land uses may involve storage and processing of hazardous or toxic materials, generation of large amounts of waste, emission of fumes, odors, and airborne particulates, and large, unsightly structures in neighborhoods.

Comparisons of Intensive and Nonintensive Uses in Zoning Patterns

Zoning codes burden low-income communities of color with intensive use designations. When one combines commercial and industrial uses, and rounds the combined figure to the whole percent, at least one-quarter of the area in each of 11 census tracts, all of them low-income, high-minority, was zoned for one of these two intensive uses, even though nearby parcels were zoned for residential uses.

On the other hand, only one high-income, low-minority census tract had any industrial zoning at all, and that industrial zoning amounted to less than 3 percent of the census tract's area. High-income, low-minority neighborhoods were the overwhelming beneficiaries of single-family residential zoning and open-space zoning. More than 75 percent of the area in each of six high-income, low-minority tracts studied was zoned for single-family residences. If open space, a country club, and a private university (with significant open space) were included with single-family residential zoning,

Several auto body shops are located within blocks of an elementary school in Old Town National City where zoning allows industrial and residential uses to be side-by-side.

The zoning disparity study, as well as other indicators of race and class disparities in planning and land use, establish a clear need for conscious, intentional efforts to incorporate environmental justice principles into land-use planning decisions.

11 of the 12 high-income, low-minority tracts had more than 75 percent of their respective areas zoned for these low-intensity land uses. The remaining tract, Costa Mesa #638.02, had more than 75 percent of the tract zoned for low-intensity land uses if the definition of low-intensity land uses includes not only single-family residences, but also a private school, a post office, a fire station, and parks, all of which are highly compatible with single-family residential uses and rarely, if ever, considered LULUs. In other words, all of the high-income, low-minority tracts had at least three-quarters of the total land uses in each tract designated as nonintensive land uses.

In contrast, the only low-income, high-minority census tract with more than 75 percent of the area zoned for single-family residential or open space uses was Pittsburgh census tract 2609.98—one tract out of nineteen. Although zoning for single-family residences or open space may preclude affordable housing needed by low-income people, the contrast in zoning patterns highlights the disparate impact of zoning designations on low-income people of color.

Lessons from the Zoning Disparity Study and Their Limits

The stark differences in zoning patterns in low-income high-minority neighborhoods and high-income low-minority neighborhoods offers two fundamental lessons for planners: 1) planning and land-use regulation, by themselves, will not necessarily produce equitable outcomes, and therefore land-use decision makers must give particular attention to the environmental justice impacts of their decision; and 2) current land-use patterns in many low-income and minority communities require a set of planning, regulatory, policy, and implementation tools that will advance the health and integrity of these communities.

However, readers should take care not to draw unsupported lessons from this zoning disparity study. The study was limited in several respects. It did not address whether race or income is more important in the uneven distribution of land-use regulation. It did not attempt to isolate the race and income variables, and statistically correlate the results to either. It did not attempt to correlate zoning patterns with the presence of any particular LULUs or environmental hazards. It is possible that a census tract with significant industrial and commercial zoning could have no hazardous waste sites, for example. It was not a longitudinal study. Thus, it did not analyze when the current zoning patterns emerged, if and how zoning patterns changed over time, and how the racial and class composition of the census tracts changed over time. The study did not attempt to identify causes of the inequitable distributions of land-use regulation.

Nonetheless, the study showed that despite the "promise" of planning and zoning to protect residential areas from intensive industrial and commercial land uses, low-income, high-minority neighborhoods are zoned for significant amounts of such uses and at higher proportions than are high-income, low-minority neighborhoods. This study, as well as other indicators of race and class disparities in planning and land use, establish a clear need for conscious, intentional efforts to incorporate environmental justice principles into land-use planning decisions.

REDISCOVERING THE PROMISE OF PLANNING IN ORDER TO PROMOTE ENVIRONMENTAL JUSTICE

The challenge—and opportunity—for land-use planners is to rediscover the promise land-use planning offers to achieve a just, healthy, and good environment for all peoples. Concern for equity and social justice has long been a core concern in urban planning (Krumholz and Forester 1990; Krumholz and Clavel 1994; Davidoff 1965; Catanese 1984). The following sections offer a number of related reasons for incorporating environmental justice into land-use planning and decision making.

To Promote Public Health and Safety

Local government authority is greatest when it aims to protect the public health and safety (*First English Evangelical Lutheran Church v. County of Los Angeles* 1989, 1361–63, 1370; Bobrowski 1995, 701). The land-use planning and regulatory functions of local government serve many valuable purposes, but the most essential purpose is the protection and promotion of public health and safety. Regulation of building design and construction prevents unsafe structures. Management of growth prevents unsafe traffic conditions from vehicle activity that exceeds the capacity of area roads. Restrictions on the location of development prevent building on unstable slopes, in flood-prone areas, over sources of public water supplies, or near sources of toxic, radioactive, or other harmful substances. Even land-use standards for open space, parks, landscaping, and the control of signage and noise serve to promote good mental health amid the stresses of urbanized environments.

Photos by Candance Rutt, PhD

Minority communities' lack of sidewalks, bike lanes, and crosswalks is one of several environmental justice issues related to transportation planning and infrastructure. These conditions pose safety hazards and discourage healthy activities like walking and biking.

Protection and promotion of public health and safety in low-income and minority communities compel incorporation of environmental justice considerations into land-use planning, regulation, and decision making. These communities face two health-and-safety problems from historic and existing land-use patterns: 1) the residents' exposure to health and safety harms, or heightened risks of harms; and 2) lack of environmental conditions that promote healthy behaviors and outlooks among residents.

With respect to harms, low-income and minority communities are home to—or closely proximate to—land uses that emit air pollutants, discharge water pollutants, use and dispose of toxic substances, spill or leak hazardous materials, create odors and noise, and generate electromagnetic fields and radiation. Many of the environmental conditions to which low-income people of color have heightened exposure are known to pose health risks (Collin 1992, 501-2; Maantay 2001; Maantay 2002). Industrial areas, with heavy truck traffic, dangerous operations, and potentially harmful materials, are hardly safe places for activities like walking, running, and playing, which typically occur in and around residential areas. In addition, certain nonindustrial land uses like liquor stores and adult establishments are known to contribute to heightened levels of crime and may be located in relatively high concentrations in areas near low-income and minority residences (e.g., Maxwell and Immergluck 1997).

Low-income and minority residential areas that lack sufficient parks, open spaces and green spaces, landscaping, public/community areas, sidewalks, visually attractive buildings, and other place-enhancing features are not healthy and safe.

Likewise, low-income and minority residential areas that lack sufficient parks, open spaces and green spaces, landscaping, public/community areas, sidewalks, visually attractive buildings, and other place-enhancing features are not healthy and safe (Harwood 2003, 25). In these environments, residents lack opportunities for exercise and physically healthy lifestyle choices (Day 2006). These environments may invite crime, vandalism, and further deterioration of the physical and social environment. Safety may be a concern. For example, low-income and minority people face a disproportionate risk of pedestrian accidents, as their communities lack adequate numbers of street lights, stop signs, and other traffic calming infrastructure (Harwood 2003). These environments can also be stressful and psychologically unhealthy, contributing to the alienation of area residents from their physical environment and community (Kahn 1999).

Land-use planning and regulation can prevent or eliminate land uses that pose health risks from low-income and minority neighborhoods and enhance facilities and resources that promote health in these neighborhoods. In addition, land-use policies can promote safe, healthy areas as sites for the development of affordable and mixed-income housing projects, schools, and other facilities serving vulnerable populations. Finally, good land-use planning and decision making evaluates the health and safety risks of all projects and plans, including analyzing cumulative and synergistic impacts, and bases decisions on the prevention and elimination of health and safety risks for all persons and groups.

To Promote Vibrant, Healthy Communities

Land-use planning and regulation improve the community's capacity to achieve its goals. Typically, members of neighborhoods have community goals that extend far beyond excluding a particular LULU from the neighborhood. They often have goals about parks and other recreational uses, open space, traffic patterns and safety, availability of grocery stores or medical facilities, maintenance of property and cleanup of nuisances, public infrastructure like streets, sidewalks, and drainage, public areas or commons, social or cultural centers, historic preservation, community identity, economic development, public transportation, and many other matters. Land-use plans contain these goals, and land-use regulations facilitate efforts to reach the goals by defining the permissible land uses in the neighborhood. Efforts to identify and achieve a "healthy" neighborhood land-use pattern are proactive, preventative strategies that will not necessarily always preclude the siting of LULUs or eliminate all environmental harms and risks, but will make both situations less likely to arise in the neighborhood with a good land-use plan than if the residents or local decision makers were to wait to react to specific proposals.

To Conserve Social Capital, Cultural Resources, and the Natural Environment.
It is now well understood that the strength and vitality of a locality or region depends on wise management of its social, cultural, and natural resources. Although there is no doubt that economies need physical capital, financial capital, and human capital, no doubt they also need the capacity of people to form networks and to cooperate for common purposes (social capital) (Fukuyama 1995, 10), rich and distinctive cultural traditions and innovations (cultural capital) (Conzo 2006; Singer and Ploetz 2002), and healthy, sustainable, functioning ecosystems (natural capital) (Daily 1997; Salzman 1997; Costanza et al. 1997). As Jane Jacobs (1961, 136–40) observed, low-income and minority neighborhoods develop organic, resilient, dynamic social networks that should not be displaced by planned development. Likewise,

TABLE 2-5. MATRIX OF LAND-USE ACTIONS INTO WHICH ENVIRONMENTAL JUSTICE CAN BE INCORPORATED

Methods of Implementing Environmental Justice Policies	Aspects of Land Use Planning and Regulation					
	Comprehensive Planning (jurisdiction-wide and neighborhood specific)	Zoning (and zoning changes)	Discretionary Permits and Negotiated Land Use Approvals	Provision of Public Infrastructure	Redevelopment	Intergovernmental Cooperation
Process (accessible, open, participatory)						
Standards (reflect environmental justice principles)						
Assessment of Current Conditions (conditions of low-income and minority communities)						
Goals and Vision (reflects goals and vision of low-income and minority communities)						
Options (alterntaives generated with input from low-income and minority communities; options include prevention and minimization of harms to these communities)						
Assessment of Impacts (assess environmental, health, social, and economic impacts on low-income people and people of color)						
Decisions (promote healthy, vibrant, low-income and minority communities; consistent with standards and goals that reflect environmental justice principles; result from open and participatory processes; prevent, eliminate, minimize, or mitigate adverse impacts on low-income people and people of color)						
Enforcement (laws, code, and permit conditions enforced in low-income and minority communities)						

low-income and minority communities have rich histories and cultural dynamics that are worth preserving (Conzo 2006; Singer and Ploetz 2002). Furthermore, environmental degradation in low-income and minority areas threatens nature's interconnected ecological systems, such as watersheds or climate, as well as a locality's environmental attractiveness to people (Spyke 2001; Hutch 2002). Conservation of social capital, cultural capital, and natural capital in low-income and minority neighborhoods by protecting them from harmful or exploitive land uses serves to strengthen and preserve the neighborhoods themselves, the city and region, and—cumulatively—society as a whole.

The reality of public participation falls short of the ideal.

To Encourage Civic Engagement and Participation, To Promote Deliberation About Good Public Policy, and To Build Democratic Institutions

The exercise of government power to plan community development, to regulate land uses, and to manage public resources is inherently public in nature. This simple observation has several implications in a representative democracy, such as the U.S., with built-in constraints on the unfettered exercise of political power:

1) Land-use planning and regulation is inherently and necessarily political (Forester 2001; Catanese 1984).

2) Government decision makers are accountable to the public for their land-use decisions, often in multiple ways.

3) Norms of open government and public access shape the process by which government entities plan and regulate land uses.

4) Input from the public—about local conditions, issues, and problems, about goals, interests, and values, and about ideas and opportunities—is critically important to expert planners, citizen-experts (e.g., planning commissioners and other appointed officials), and elected representative of the public (e.g., city or county council members and mayors).

5) Some level of meaningful public engagement and participation in planning and regulatory processes is necessary for the legitimacy and long-run efficacy of government actions. Having a voice in decision-making processes affects perceptions about the fairness of the outcomes (Folger 1977) and about the legitimacy of the institution itself (Hirschman 1970).

6) Land-use planning and regulatory processes play an important role in building democratic institutions, contributing to public learning and deliberation, and enhancing the quality and scope of the public's engagement with civic and community life generally (Forester 2001). This is true not only for the local political processes, but also for the entire society, especially because: a) local government is the one closest (and arguably most accessible) to people, and b) land-use decisions have direct, concrete, and particularly salient impacts on people and their day-to-day lives.

The reality of public participation falls short of the ideal, though. There is a strain of planning theory that questions whether non-expert local residents have the analytical skills and information to grasp complex, technical, long-term land-use issues or can (or will) overcome their individual self-interest to seek goals and policies that advance the public good (Lucy 1988, 147–48). In fact, some planners may resist or resent the involvement of community residents, and even of elected or appointed officials. However, a greater amount of frustration by planners likely comes from the imperfections of the political process. Despite democratic norms or ideals, many voices or interests are left out of land-use decision-making processes. Power is unevenly distributed, and special interests—or at least, competition and conflict among special interests—dominate land-use politics. Limited time, information, human cognition, commitment, and trust reduce potential public participation and hamper local officials from reaching decisions that reflect the community's will or goals.

Improvements in the practice of local land-use democracy are possible, though, even if achievement of a "civic republican utopia" is not. An extensive planning literature calls for democratic and participatory planning processes that engage and involve an informed, deliberative public (Burke 1979; Forester 2001; Arnstein 1969; Fagence 1977). Moreover, the very principles of environmental justice call for enhanced participation by all peoples, including the often-ignored low-income communities of color. Environmental justice principles also call for mechanisms to involve community residents, to receive and consider their input about problems and goals, and to enhance government's openness and accountability to the public. They seek to strengthen democratic institutions, to empower community residents, and to build the capacity of community residents to shape the direction of their own neighborhoods. Public faith in local governmental institutions will increase as the breadth and depth of public involvement increases. A locality's systematic and sustained attention to environmental justice will contribute to a more dynamic and resilient quality of public life.

Environmental justice principles call for mechanisms to involve community residents, to receive and consider their input about problems and goals, and to enhance government's openness and accountability to the public.

To Promote Efficiency and Certainty, and To Reduce Conflict-Related Costs

Local land-use planning and regulation create greater certainty about what land uses will or will not be allowed in a neighborhood than does reliance on federal and state environmental regulation, common law doctrines of nuisance or trespass, or civil rights litigation.

When local land-use regulations allow LULUs, either by right or conditionally, neighborhood residents face uncertainty about whether their neighborhood will be the object of an LULU siting proposal (or a proposal to site another LULU in their neighborhood if they already have one or more), and once a proposal has been made, whether they will be successful in defeating the proposal. Similarly, the property owner, developer, or business operator faces uncertainty about whether local residents will attempt to defeat the project as inappropriate for the neighborhood even though the local land-use regulations permit it and the owner or operator has invested significant amounts in that specific site proposal. Both sides have significant economic costs (i.e., inefficiency), psychological costs (i.e., anxiety), and relational costs (i.e., suspicion and animosity) resulting from uncertainty about the propriety of the LULU in the neighborhood. The potential for costly and bitter conflicts, including litigation, is quite high.

If local residents have been involved in the land-use planning and development of regulations for their neighborhood, however, and have carefully identified what uses are appropriate for what areas of their neighborhood, the level of certainty increases substantially. Proponents of LULUs may nonetheless seek amendments to or relief from applicable land-use prohibitions, and neighborhood residents may nonetheless oppose LULUs permitted by the regulations. Disputes and litigation are still possibilities. But in most circumstances, the content of the land-use plans and regulations, when developed with meaningful neighborhood participation, provide generally reliable information on which both sides can make decisions. This information fosters efficiency, comfort, and trust.

To Provide Opportunities for Coordinated, Collaborative, and Proactive Solutions To Multiple, Complex, Interconnected Problems

Good planning and land-use practices address problems in their entirety, taking a coordinated and integrated approach to problems with multiple parts or to multiple, interrelated problems. They also seek solution-generating collaboration among all of the groups affected by land-use problems or challenges. And, furthermore, they take a proactive approach to land-use issues, seeking to avoid problems before they arise or to solve, rather than merely remedy, problems. Of course, a totally comprehensive planning or decision-making process is not likely achievable for complex problems (Lindblom 1959), and plenty of land-use conflicts exist that present enormous obstacles to cooperation among all the stakeholders (Forester 2001). Nonetheless, incorporating environmental justice principles and policies into land-use planning and decision making measurably improves the public problem-solving process. Environmental conditions in low-income and minority communities are then recognized as being interconnected with other problems in those communities, such as land-use patterns, health, economic conditions, crime, and housing, as well as being interconnected with other land-use, environmental, and social problems throughout the locality or region, such as growth and development patterns, the jobs-housing balance, regional air quality, public health care costs, and the like. Involving low-income communities and communities of color as important and necessary stakeholders in solving land-use problems and shaping land-use policies creates more opportunities for cooperative outcomes than does ignoring these communities, which then must turn to political protest, decision-specific oppositional tactics, and litigation to seek fair outcomes. Moreover, future environmental injustices can be avoided by a proactive approach to equitable land-use planning and regulation.

Incorporating environmental justice principles and policies into land-use planning and decision making measurably improves the public problem-solving process.

To Promote a Proactive, Prospective, Problem-Solving Model of Addressing Environmental Justice Issues, Rather than a Reactive, Oppositional Model

Many grassroots environmental justice struggles have involved legal, political, and sociocultural opposition to existing and proposed LULUs and other sources of pollution or environmental harms. This might be called the ***oppositional model of environmental justice***. However, a ***planning model of environmental justice*** is increasingly developing within the environmental justice community, as low-income neighborhoods of color seek to define and protect their communities through land-use planning and regulation. Neighborhood residents who engage in land-use planning and develop proposed land-use regulations for their neighborhood are proactively seeking to prevent LULUs or environmental harms before the siting process ever begins. Furthermore, they are defining not only what they do not want in their neighborhood but also what they do want.

The planning and opposition models of environmental justice share some characteristics. Both are largely concerned with questions of fairness and goals of achieving safe and healthy communities. Both attempt to prevent environmental hazards and LULUs in low-income and minority neighborhoods, albeit in different ways. And both are struggles for grassroots participation in policymaking and in political, economic, and legal decisions that affect these neighborhoods.

The models also differ in some important ways. In the opposition model, grassroots activists react to existing LULUs or proposed sitings. In many cases, they may seek remedies for past or ongoing harms or government and corporate decisions that pose the risk of harm. Thus, the opposition model is largely reactive, retrospective, and remedial, although perhaps necessarily so. In the planning model, local residents develop land-use plans and regulations that either address broader problems than a single LULU or reflect goals for future land-use patterns in the neighborhood. To some extent, these plans and regulations capture an element of the community's self-identity (e.g., a high-density community of affordable housing; a historic neighborhood of single-family residences and small retail businesses; a neighborhood of single- and multi-family housing with many small parks and playgrounds and few through-streets; an area in which industrial activities remain on the east side of the river). These plans and regulations also are in place to govern future land-use decisions, including proposals for LULU sitings. In these ways, the planning model is proactive, prospective, and visionary.

Opponents of existing or proposed LULUs often are political outsiders, entering the decision-making process after relationships have been established between the facility owner or operator and government officials. Theirs is the struggle of people without power who are taking on and fighting established exercises of power. Some environmental justice activists reject government decision making, economic markets, and the legal system as inherently subordinating and victimizing the poor and minorities. In many ways, low-income people and people of color who seek to influence land-use planning and regulation start out similarly struggling against the powerful. Their goal, though, is to exercise power within the existing land-use regulatory system. They want to be participants in the process, empowered by their definition of land-use goals and what they hope will be the successful implementation of these goals through zoning and other regulations. They want to be participants at the land-use negotiating table in matters that concern them, along with government officials, developers, property owners, environmentalists, and other interested people and groups. They want to serve on advisory boards, zoning commissions and boards of appeal, city councils, and other decision-making bodies. Finally, the opposition model identifies and seeks to exclude harmful activities and LULUs. The planning model identifies and seeks to allow (i.e., include) desirable land uses. The contrasts between these two models are summarized in Table 2-2.

TABLE 2-2. CHARACTERISTICS OF TWO MODELS OF
ENVIRONMENTAL JUSTICE

Opposition Model	Planning Model
Reactive	Proactive
Retrospective	Prospective
Remedial	Visionary
Outsiders	Participants
Fighting power	Exercising power
Subordinated	Empowered
Victims	Decision makers
Exclusive	Inclusive

Source: Arnold 1998, 96

CHAPTER 3

Comprehensive Planning
and Environmental Justice

E nvironmental justice concerns often arise in a particular community in the context of a specific land-use problem, infrastructure failure, or permit decision. As a result, both planners and local officials may be tempted to treat environmental justice concerns in an ad hoc, piecemeal, fragmented manner. Limited planning resources may seem to necessitate a reactive strategy. A proactive strategy, though, is critical.

Planning for environmental justice, or equitable planning, is characterized by 18 principles that can be incorporated into any local planning process.

At a minimum, a locality's comprehensive plan and all its plans and planning processes (e.g., area plans; see the discussion below) should specifically and identifiably incorporate environmental justice principles, which are discussed in the next section of this chapter. Zoning and project-specific decisions based on environmental justice considerations could be struck down if they are inconsistent with the content of the comprehensive plan and its elements.

The opportunities to incorporate environmental justice into plans are more abundant than might be initially obvious. Incorporating them should be a systematic, regular part of drafting and updating plans. These planning opportunities range from revisions or updates to general comprehensive planning documents (see the sidebar), development of policies to address specific planning problems or issues, initiation of district- or neighborhood-based planning programs, preparation of proposals for housing, community development, or distribution of transportation funds, discussion of new ideas about planning (e.g., smart growth principles), amendments to the zoning code, and feedback about plans and policies in light of specific land-use decisions.

ENVIRONMENTAL JUSTICE PLANNING PRINCIPLES

Planning for environmental justice, or equitable planning, is characterized by 18 principles that can be incorporated into any local planning process:

1. Adopt plans, policies, and regulations that are fair and achieve a healthy environment, vibrant community, and good quality of life for all peoples

2. Achieve widespread participation of all affected persons

3. Implement a vision that empowers community residents

4. Perform environmental justice audits

5. Assess and analyze environmental and health risks from existing and proposed land uses

6. Protect people from incompatible land uses, especially industrial and intensive commercial uses and uses that pose significant risks to human health and safety

7. Locate housing, schools, and facilities caring for vulnerable people (e.g., ill, elderly, children) in areas that are not proximate to industrial facilities, contaminated sites, or other land uses that pose significant risks to human health and safety

8. Plan primarily for pollution prevention and elimination, and secondarily for pollution containment and mitigation

9. Preserve diverse cultural assets in the community

10. Provide and maintain equal and adequate services and infrastructure

11. Engage in specific district planning in low-income and minority neighborhoods

12. Provide a mix of affordable housing options adequate to meet the locality's share of the regional need or the specific needs in the locality, whichever is greater

13. Promote development and land uses that provide economic opportunities to low-income and minority residents, including living wage jobs, skill development and training, and business creation and ownership opportunities

14. Clean up and redevelop brownfields with primary emphases on area-resident and end-user health and safety, and the social and economic health of the surrounding neighborhood

CALIFORNIA'S GENERAL PLAN GUIDELINES ON ENVIRONMENTAL JUSTICE

The State of California has a policy of encouraging its cities and counties to incorporate environmental justice into their general plans. A state statute requires each city and county to adopt and periodically update a formal, written comprehensive plan, known as a "general plan," containing seven mandatory and internally consistent elements: land use, housing, circulation, conservation, open space, safety, and noise. All land-use decisions must be consistent with the general plan.

In 2001, the California Legislature directed the Governor's Office of Planning and Research (OPR) to include environmental justice as an optional consideration in its *General Plan Guidelines*, which guide local governments in preparing and revising their general plans. Specifically, OPR was to identify methods for local governments to address:

- the equitable distribution of new public facilities and services that enhance community quality of life;

- the location of industrial facilities and land uses that pose a significant hazard to human health and safety to avoid overconcentration of theses uses near schools or residences would be avoided, and conversely, to locate new schools and residences to avoid proximity to these intensive land uses; and

- ways to promote more livable communities by expanding opportunities for transit-oriented development. (OPR 2003, 23)

The General Plan Guidelines now contain a chapter devoted to sustainable development and environmental justice, in which environmental justice planning goals are expressly linked to concepts of sustainable development, transit-oriented development, and the jobs/housing balance.

The Guidelines mirror many of the planning concepts contained in this PAS Report. Their recommended planning methods and principles for environmental justice include:

- *Integration*. Ideally integrate environmental justice into all of the elements of the general plan, instead of adopting a separate environmental justice element.

- *Public Participation*. Pursue robust public participation strategies because environmental injustice is not only geographically inequitable in the distribution of undesirable land uses but also procedurally inequitable inasmuch as some segments of the community face barriers to participation. "Participation plans should incorporate strategies to overcome linguistic, institutional, cultural, economic, and historic barriers to effective participation." (OPR 2003, 23) These strategies include: outreach and advertising; communications in the local dominant languages; holding meetings at times and places accessible to the affected communities; collaboration with stakeholders; transparency with both process and information; gathering community concerns through surveys and meetings; and use of participatory workshops and community-based planning advisory groups.

- *Jobs/Housing Balance*. Carefully plan for the location, intensity, and nature of jobs and housing to locate people's jobs and residences in close enough proximity to one another such that overall vehicle trips and miles are reduced. Methods include higher-density housing near centers of employment; infill development; affordable housing development programs; recruitment of businesses that will use the local workforce; local workforce training; a strong telecommunications infrastructure; and mass transit, alternative transportation modes, and pedestrian-friendly development.

- *Compatibility*. While welcoming mixed-use development for sustainability, segregate incompatible uses, particularly the location of residential and school uses in proximity to: 1) industrial uses; 2) uses that generate substances posing a significant hazard to human health and safety; 3) intensive agricultural uses; 4) major thoroughfares, such as highways; and 5) resource extraction and production activities, such as mining or oil and gas wells.

- *Information*. Gather socioeconomic data, community-profile data, and land-use distribution data; analyze that data using a geographic information system (GIS).

- *Public Facilities*. Achieve an equitable distribution of public facilities throughout the community, measured by the number and size of facilities and the residents' access to these facilities. Locate facilities within walking distance, along transit corridors, and in multifunction urban centers (depending on the type of facility), and make public facilities open and accessible to all local residents.

- *Industrial Facilties*. Prevent or reduce the proximity of industrial facilities to residences and schools, and the overconcentration of industrial facilities in or near neighborhoods. Methods include: buffer zones between industrial and residential land uses; policies to consider and mitigate environmental impacts in conditional project-siting decisions; caps on the number of certain facilities within proximity to one another; and rezoning mixed residential/industrial areas to prevent new or expanded industrial uses.

- *New Residential Facilities and Schools*. Reject new residential development in areas with high concentrations of industrial uses. Create buffer zones between existing industrial areas and new residential areas. Identify appropriate areas for new schools and housing to avoid exposure to industrial land uses.

- *Transit-Oriented Development*. Pursue transit-oriented development, which is defined as "moderate- to high-density development located within an easy walk of a major transit stop, generally with a mix of residential, employment, and shopping opportunities." (OPR 2003, 28) Suggesting a variety of planning, regulatory, and incentive-based tools to promote transit-oriented development, OPR notes the benefits of reduced air pollution and community exposure to air pollution, the promotion of healthy activities (e.g., walking and cycling), and community access to jobs, retail uses, businesses, and public facilities.

Sources: California Government Code, Sections 65300 et seq.; *Citizens of Goleta Valley v. Board of Supervisors* 1990; California Governor's Office of Planning and Research 2003.

15. Prevent the displacement or expulsion of local residents by gentrification, redevelopment, new development, or brownfield remediation

16. Pursue transportation policies that reduce automobile usage and overall vehicular emissions, distribute air quality impacts evenly throughout the region, and provide effective transportation options to low-income and working class people

17. Plan for open spaces, green/natural spaces, recreational spaces, civic spaces, market spaces, and artistic and cultural spaces in all areas of the locality

18. Ensure that zoning, other land-use and environmental regulations, public projects and expenditures, and permit decisions are consistent with plans incorporating environmental justice principles

Environmental justice (or equitable) planning defines land-use compatibility and incompatibility based on the environmental and social impacts of land uses.

These principles cluster around five conceptual categories: participation, land-use compatibility, pollution prevention and elimination, community preservation, and infrastructure development. Each category is discussed below.

Participation

All land-use decisions, including comprehensive and specific planning, zoning, land-use permit decisions, redevelopment, and infrastructure development, result from full and meaningful participation by the people who will be affected by the plans and from processes that are open, accessible, and comprehensible. Plans reflect the vision of local residents about their community and future, resulting from the integration of the expertise of planners and other professionals, the leadership of political and community leaders, and the "grassroots" participation of community residents. Planning and other land-use decision making occur at times, at locations, and in languages that involve a wide range of people in the community. Planning and land-use decision making empower the people and communities affected by land-use decisions. Chapter 5 provides specific ideas about creating, enhancing, and welcoming opportunities for full and meaningful participation by all affected peoples.

Land-Use Compatibility

Environmental justice (or equitable) planning defines land-use compatibility and incompatibility based on the environmental and social impacts of land uses.

A core component of comprehensive planning is the identification of compatible land uses (i.e., permissible or desired uses) and incompatible land uses (i.e., prohibited or discouraged uses). Traditional Euclidean zoning defines compatibility by categories and subcategories of land uses (e.g., industrial, commercial, residential, and agricultural). Ad hoc planning and piecemeal zoning practices define compatibility on a project-by-project basis, often with emphasis on the economic and social impact of the proposed project and the political power of surrounding property owners or occupants. Contemporary form-based planning and zoning define compatibility by design standards, attempting to encourage or achieve mixes of uses appropriate to a particular context.

In contrast, planning and zoning for environmental justice requires attention to the health and safety of low-income people and people of color, as well as to the vitality of low-income and minority neighborhoods. Land-use plans and decisions should segregate land uses that pose substantial health risks, produce significant pollutants or hazards, or threaten the safety of neighbors from residential land uses, schools and day care facilities, religious land uses, parks and playgrounds, medical care facilities, and similar places of potential prolonged exposure to unprotected people. Likewise, land-use plans and decisions should locate housing, schools, and facilities caring for vulnerable people (e.g., ill, elderly, children) in areas not proximate to industrial facilities, contaminated sites, or other land uses that pose significant risks to human health and safety.

For example, environmental justice compatibility analysis would preclude plans that identify multifamily housing as an appropriate "buffer" land-use to create a transition from single-family neighborhoods to industrial districts. Likewise, from an environmental justice analysis, plans should not designate land immediately adjacent to freeways as appropriate sites for multifamily housing, due to the air quality impacts of freeway traffic, as well as noise and visual impacts. Environmental justice compatibility for mixed-use projects calls for careful attention to the environmental and social impacts of the specific uses in the project. For example, affordable condominiums or apartments located above a business that sells books is quite different than over a facility that prints books. A mixed-use project that combines residential units with retail stores and restaurants is mixing compatible uses if the commercial uses are grocery stores and family restaurants but is mixing incompatible uses if they are liquor stores and exotic dance clubs.

Pollution Prevention and Elimination

An important environmental justice component of land-use plans and policies is a primary emphasis on pollution prevention and elimination, with a secondary emphasis on pollution containment and mitigation. Planning for environmental justice recognizes that an individual facility's plans for pollution containment or mitigation will often not be adequate given economic- and optimism-driven tendencies to underestimate risk, unintended accidents and mistakes, regulatory slippages and gaps, regulatory underenforcement, poorly understood synergistic impacts of pollutants, and the cumulative impacts of pollutants, particularly in low-income and minority communities. These concepts apply not only to waste facilities and industrial land uses but also to transportation facilities, traffic patterns, brownfield cleanup projects, and stormwater runoff patterns, among others.

Community Preservation

Planning for environmental justice takes affirmative steps to protect and to preserve low-income and minority communities. These communities typically have rich histories and cultural traditions, strong and functional networks, clear community identity and sense of place, and community-specific assets and resources. To be sure, many of these communities also face or suffer harms, degradation, stresses, and problems that threaten the community's identity and assets. Nonetheless, principles of equity (fair treatment of the communities' residents), humanity (respect for the integrity and inherent value of these communities), and efficiency (conservation of communities that are assets to the larger metropolitan area or region) call for policies to strengthen and sustain those communities, while preventing or eliminating threats to them.

Infrastructure Development

Local governments plan for, develop, and maintain public facilities, services, and infrastructure that are adequate to support the needs of the specific neighborhood or area and that are equitable with respect to all neighborhoods and areas within the local jurisdiction. Environmental and land-use

An important environmental justice component of land-use plans and policies is a primary emphasis on pollution prevention and elimination, with a secondary emphasis on pollution containment and mitigation.

Environmental justice principles and smart growth principles can be integrated in several different ways, yet smart growth policies may need a few clarifications or modifications so that all people are treated equitably.

injustices in minority and low-income communities are defined not only by harms, such as pollution, LULUs, and industrial land uses, but also by lack of benefits, such as parks and open space (see Figure 3-1), public transportation options, up-to-date utility services (including water supply distribution systems, sewer systems, and stormwater drainage systems), affordable housing, healthy streams and rivers, community centers and recreational facilities, and well-landscaped and well-maintained streets and sidewalks. Good land-use planning gives priority to remedying inadequacies and inequities in public facilities and services and to enhancing and sustaining the neighborhood's physical and social infrastructure.

ENVIRONMENTAL JUSTICE AND SMART GROWTH

In thinking about the planning principles of environmental justice, planners may wish to consider the relationship between environmental justice and smart growth. Environmental justice principles and smart growth principles can be integrated in several different ways, yet smart growth policies may need a few clarifications or modifications so that all people are treated equitably.

Planning experts have given considerable attention to the relationships between smart growth and environmental justice (Hutch 2002; Kushner 2002-2003; Collin and Collin 2001; National Governors' Association Center for Best Practices 2001). The American Planning Association's Policy Guide on Smart Growth (2002) identifies "social equity and community building" as one of its five categories of policy positions on smart growth. It states that smart growth planning principles promote equitable processes and equitable resources distribution, as well as participation by "a diversity of voices" in community planning and implementation. This policy is available at www.planning.org/policyguides/smartgrowth.htm and is summarized in Appendix B of this PAS Report.

Even more than smart growth concepts, concepts of sustainable development make systematic links between environmentally sustainable practices and social equity and justice. Despite critiques that the concept of sustainability is too amorphous, capable of meaning all things to all people, and attempting to link incompatible elements (economic growth and development; social equity and justice; and environmental protection), planning and environmental experts believe that sustainability concepts can offer a new vision of environmental and land-use policy that is both ecologically sustainable and socially just (Agyeman 2005; Collin and Collin 2001; American Planning Association 2000; Campbell 1996). The American Planning Association's Policy Guide on Sustainability (2000) contains several policy proposals that promote both smart growth and environmental justice principles. This policy is available at www.planning.org/policyguides/sustainability.htm and is summarized in Appendix B of this PAS Report.

Several specific components of a smart growth agenda clearly advance environmental justice norms and goals. These components include the following:

- Decreased sprawl and an end to disinvestment in the central city, both of which contribute to racial and economic segregation and the deprivation of resources for central-city residents and services

- Revitalization of central cities with sustainable urban neighborhoods and vibrant urban environments

- Remediation and revitalization of brownfields: removing contamination from areas where low-income and minority people live and making good use of unused or underused properties that have already been developed, have contributed to neighborhood decline when contaminated and under-

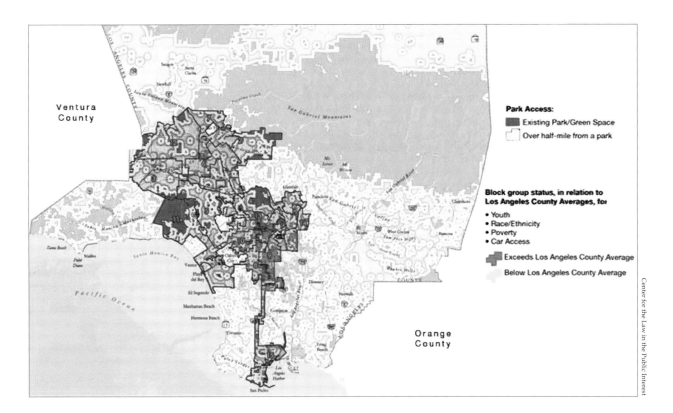

used, and can contribute to job creation, housing supply, or provision of other community land-use needs if remediated and reused

● Attention to the environmental impacts of land development and to ways preventing, eliminating, minimizing, and mitigating these impacts

● Promotion of green building standards and environmentally sustainable methods of development

● Transit-oriented development that reduces traffic, improves air quality (especially in "hot spots" in low-income and minority communities), connects people's homes and their jobs, improves the means for pedestrian activity in cities, and increases mass transportation options for urban residents

● Development standards that promote healthy lifestyles for all area residents

● Creation of mixed-income housing throughout the metropolitan area

● Increased supply of affordable housing in central cities

● Investment in projects that increase and enhance open space, natural areas, recreational facilities, and community/public gathering spaces (e.g., squares, plazas, mixed-use environments) in ways that enhance community life, connect people to nature, and celebrate a sense of place

● Sustainable land-use practices and restoration activities for urban waters and watersheds

● Empowered local communities through active participation in land-use planning and decision making (American Planning Association 2002; National Governors' Association Center for Best Practices 2001; Hutch 2002; Arnold 2005; Garcia and Flores 2005).

Figure 3-1. Park Access for Children of Color Living in Poverty with No Access to a Car

Planners, government leaders, and smart growth advocates should be cautious about the several ways a smart growth agenda might be inconsistent with environmental justice principles unless the agenda is carefully circumscribed.

Clearly, the realm of shared values between smart growth and environmental justice is extensive. At the same time, though, planners, government leaders, and smart growth advocates should be cautious about the several ways a smart growth agenda might be inconsistent with environmental justice principles unless the agenda is carefully circumscribed. Consider the following potential problems:

1. The selective misuse of smart growth policies to restrict development of industrial parks and multifamily housing on the urban/suburban edge can force these two uses into the central cities in close proximity to one another.

2. New Urbanist developments, community-promoting spaces, sustainable natural areas, and pedestrian- and transit-oriented development can become high-priced luxuries if efforts are not made to accommodate affordable housing.

3. Related to this point, the revitalization of central cities and channeling of housing and development to already developed urban areas can and has resulted in gentrification, displacement of current central-city residents, and loss of community identity for low-income and minority neighborhoods.

4. Infill development can occur in open space areas in central cities, further adding to the amount of impervious cover and loss of open space in those areas; infill might be better focused on reusing already developed parcels of land.

5. The new smart growth emphasis on mixed-use development and form-based zoning emphasizes visual and design elements of land-use districts (Local Government Commission 2004; Congress for New Urbanism 2004) to ignore the necessity of segregating certain types of incompatible uses, such as industrial and residential uses. Although the concept of locating housing close to other destinations is laudable in the abstract, we should not forget that one of the reasons for the existing exposure of low-income people of color to environmental harms and risks is the historic location of workers' housing near the pollution-generating industries in which they worked.

6. Finally, smart growth principles call for regional planning. There is reason for caution, though, because regional planning may impede the participation and empowerment of low-income and minority communities that may have less access to the process, fewer opportunities to participate, and diminished political strength relative to the vast number of communities involved in regional land-use planning and regulation. The voices of low-income and minority neighborhoods could get lost in the broader scale of decision making.

Overall, the answer is not to lose sight of environmental justice as offering an important set of planning principles and imperatives consistent with smart growth principles, yet distinct from smart growth principles. Each set of principles has an important role to play in new modes of land-use planning and practice.

AREA-SPECIFIC PLANNING IN AND BY LOW-INCOME MINORITY COMMUNITIES

Much of the work of planning for environmental justice and equitable land-use occurs in area-specific or neighborhood planning. The specific geographic focus of a planning process might be: 1) a neighborhood; 2) a district or sector, which are terms for a collection of neighborhoods or an identifiable area larger than any particular neighborhood; or 2) a corridor, which is an area along a major road or highway (Kelly and Becker 2000, 323–38).

The participation of community residents is especially important in area planning, especially in areas with low-income, minority populations because:

- environmental justice problems and harms are often particular to certain areas of a community, especially low-income and minority neighborhoods;

- when the focus is their neighborhood or an identifiable set of related neighborhoods, residents of low-income, minority communities are more likely to get involved than when the focus is the entire local jurisdiction or the region from which they may already feel disenfranchised (Kelly and Becker 2000, 323); and

- area-specific plans allow for the development of objectives, action items, and outcomes with greater specificity and detail than broader communitywide plans (Kelly and Becker 2000, 323).

Nonetheless, neighborhood and district plans "should be done in the context of a communitywide plan [and] . . . should support the broader needs of the community and region" (American Planning Association 1998).

Selection of Areas and Issues for Planning

Planning that addresses environmental justice issues and achieves equitable land-use patterns will focus on those neighborhoods, districts, and corridors subject to, or at risk of, disproportionate environmental impacts, land-use burdens, or marginal-to-poor infrastructure. More specifically, factors in selecting a geographic area and defining its boundaries include the following:

1) The predictable vulnerability of the area's residents to existing or new environmental burdens and intensive land uses, especially given demographic characteristics such as race, ethnicity, income levels, and poverty levels

2) The identifiability of the area as a neighborhood based on commonly recognized community identity, sociocultural features, and physical boundaries

3) The planning relevance of neighboring facilities and land uses likely to have a significant impact on an identifiable neighborhood or district (i.e., include these "border areas" in the planning area)

Comprehensive planning for a particular neighborhood or district is preferred to issue-specific planning because a broad, integrated, multi-issue plan for an area's development and future will more likely produce coordinated responses to discrete but related problems and result in a compelling vision that has community support. From time to time, however, the development of plans to address specific issues or sets of issues in a particular neighborhood or district may be necessary. The urgent need to address the issues quickly, limited staff or financial resources for community planning, or the existence of a functional neighborhood plan that is merely lacking attention to one issue (or a few related issues) may justify not only area-specific planning, but also issue-specific planning. In these cases, planners will need to define the issues as precisely and clearly as possible, and to communicate effectively both the issues and the limited scope of the planning process to the participants.

Planning that addresses environmental justice issues and achieves equitable land-use patterns will focus on those neighborhoods, districts, and corridors subject to, or at risk of, disproportionate environmental impacts, land-use burdens, or marginal-to-poor infrastructure.

Environmental injustice is not only about the environmental conditions experienced by low-income people and people of color, but also about their lack of power over land-use and environmental decisions and practices in their communities.

Public Participation

Neighborhood or district planning is a partnership between planners and government officials on one hand, and community residents and property owners on the other. Involvement of area residents in meaningful and empowering ways in all stages of planning for their neighborhood(s) is critical for several reasons. Environmental injustice is not only about the environmental conditions experienced by low-income people and people of color, but also about their lack of power over land-use and environmental decisions and practices in their communities. Planning enjoys its greatest legitimacy in a representative democracy when the public participates extensively and effectively. Plans are more likely to enjoy success when the affected people share in shaping the plan's vision for what the area could or should be, determining policy objectives and action items for the area's development and growth, and making or defining the place—and sense of place—and building or strengthening the community as the plan is implemented. Consensus from a collaborative process reduces the risk of long-term conflict that results in economic, political, and social costs, as well as policy failure. "Research conducted by the American Planning Association and other groups has shown that the best neighborhood plans are developed by informed residents collaborating with decision makers, service providers, and business leaders in a process designed and facilitated by neighborhood planners" (American Planning Association 1998).

Chapter 5 of this PAS Report contains a variety of ideas about enhancing community participation in planning. Several additional techniques in conducting neighborhood planning, though, may maximize the scope and quality of public participation:

- Hold meetings at times and in locations where the maximum number of participants can come.

- Facilitate the planning process and provide structure, without controlling or directing it. Be a partner, neither a passive scribe nor an assertive director. Do not hesitate to contribute planning expertise (i.e., information, analytical skills and tools, experience, ideas) but recognize expertise is not a substitute for community values and consensus, and community residents may be skeptical about or opposed to reliance on scientific and technical expertise.

- Listen to residents' goals, aspirations, needs, and concerns prior to generating options and alternatives.

- Keep presentations to residents short and informative, while providing ample time for discussion and participant input.

- Write down participants' comments in their own words; translation into the language of planning experts at the time participants' comments are recorded on paper contributes to misunderstandings and feelings of not being understood; relationships between the residents' framing of issues and the planners' framing of issues can be addressed at other points in the process, not in the recording of participant input itself.

- Make sure that all participants' voices are heard and try to prevent one person or a few people from dominating the process. Set ground rules that include participants' listening to one another, keeping focused on the issues (not personalities), and being respectful of one another's views.

- Use the active listening technique of recaps by which the facilitator reflects back to the participants what he or she understands that he or she has heard, giving the participants an opportunity to confirm, clarify, or

Environmental justice planning includes community-based visioning and planning processes.

supplement the facilitator's understandings of their comments. Recaps contribute to a sense of momentum and accomplishment that will keep people engaged and involved.

- Identify and use tasks that keep participants involved and engaged. Be organized to make effective use of time. Use break-out groups to give participants more opportunities to voice their perspectives in small group settings.

One of the more vexing issues in area-specific planning is whether, how, and when to involve owners and operators of businesses and industrial facilities in the area. As stakeholders, they are entitled to participate. They are likely to express their interests, and if they are excluded from the process, it is likely to derail the plans developed. In addition, dialogue between the residents and area businesses and industries is an important component of addressing and solving environmental and land-use problems in the area. Although community residents may like to take control of land-use practices in their communities, true empowerment means negotiating and problem solving with the businesses and industries in their areas, not merely engaging in conflict with them. A good planning process for areas characterized by environmental justice issues, however, will have at least some meetings just for community residents and planners, usually early in the process. Community residents will want to talk among themselves and with city officials, without having the industries or businesses that concern them dominate the discussion or foreclose their input and ideas. Residents may feel that businesses and industries have had influence over land-use decisions in their communities, while they have not. The early and vocal presence of business and industry interests may create a perception the planning process is already stacked in business's favor and community residents are there only to be coopted or to lend a false sense of legitimacy to the process. Some people may be hesitant to voice their concerns about existing facilities if the owners or managers of those facilities are present at the first time those concerns are being aired. At the same time, though, planners and community residents should talk up-front about future planning meetings that involve business and industry interests.

A Five-Stage Process

Although many good neighborhood or district planning processes exist (e.g., Kaiser et al. 1995; Kelly and Becker 2000; Martz 1995), planners might consider the following five-stage process in planning for low-income and minority areas where environmental justice issues prevail. The sidebar about community planning in the Dudley Street neighborhood of Boston illustrates how each of these stages may be present in a neighborhood-based planning process, even if not in the systematic, linear manner proposed in this PAS Report.

Stage 1: Building relationships and defining the process. The participants, including planners and other officials, need to know one another and have at least enough trust in one another to engage in the common enterprise of planning the area's future. The planners, community residents, and others must have a working relationship that facilitates communication and the work of planning, and allows the process to move forward. In addition, the participants must understand and have some degree of consent to the scope of the planning process (geographic scope and scope of issues), the nature of the planning process itself, and the procedures and standards for the local government to adopt and implement a plan.

Stage 2: Assessing current conditions. Prior to defining needs, goals, options, and alternatives, all participants need to understand the current physical, social, and economic conditions of the area or neighborhood. The environmental justice audit discussed below can serve as a useful guide in gathering and presenting this information. Planning staff and other government officials will likely be the source of much of the information. A particularly effective tool for community participants and an informative tool for planners, however, is to involve community residents in gathering and sharing information about current conditions. For example, having community residents, or even area schoolchildren, participate in doing a land-use inventory helps to shape an understanding of issues, needs, and opportunities, as well as providing useful information to expert planners.

Presenting data about current conditions in a variety of formats will help planning participants understand and make good use of the information. These formats include maps identifying land uses, facilities, and features of the community; charts, graphs, and tables presenting statistics; photographs of streetscapes, buildings, and similar physical conditions in the area; computer simulations, slide shows, histories, and narrative descriptions. Geographic information systems (GIS) software tools are especially helpful in analyzing and presenting data. The data should include environmental conditions and social conditions, as well as land-use conditions. It should include community assets and resources, as well as problems and limitations.

Stage 3: Identifying needs and goals. Discussion and identification of the community's needs and goals for the area should precede the process of considering options and alternatives for the area's future. It is tempting for planning processes to be driven by options and alternatives, with goals and needs being defined by concrete ideas about what could be achieved. For example, if a land-use survey of the community were to show four contiguous parcels of empty, unused warehouses, it might seem logical to proceed to a series of possible land-use options for a single large unit of land composed of the four parcels—options that would be appropriate to the surrounding land uses, area infrastructure, and market conditions. As a result of this alternatives-first approach, the community residents might define one of the plan's goals on the basis of their most preferred use for the site(s) in question. However, alternatives-driven goal-setting forecloses the identification of goals that do not correspond to pre-identified options

The Dudley Street Initiative resulted in the construction of 2,000 affordable homes, the creation of an urban village, and the end to the displacement of neighborhood residents. See the sidebar on page 43 for a more detailed discussion.

DUDLEY STREET NEIGHBORHOOD PLANNING IN BOSTON

The Dudley Street Neighborhood Initiative (DSNI) in the Roxbury area of Boston represents effective neighborhood-based planning with strong community participation. Roxbury is a 4.2-square-mile area with a poverty rate approaching one-third of all residents. Less than one-quarter of all housing was owner-occupied in the early 1990s, and the number of area businesses had declined by 80 percent from 1950 to 1980. The Dudley Street Neighborhood is a 1.5-square-mile "multilingual low-income community of color" within the heart of Roxbury (Alves et al. 1995, 736). The combination of white flight, redlining, insurance-related arsons, illegal trash dumping, absentee landowners, disinvestment, land-use decisions, and neglect left the Dudley Street Neighborhood in physical decline. The neighborhood was home to more than 1,000 vacant lots, more than 64 percent of Boston's solid waste storage and transfer facilities, 54 state-listed hazardous waste sites, and 15 bus and truck depots serving more than 1,000 diesel vehicles.

Dudley Street Neighborhood residents responded to these conditions by creating neighborhood-driven plans and redevelopment projects. This planning process began in conflict. The Boston Redevelopment Authority (BRA), the city's land planning agency, had proposed an area redevelopment plan, which residents opposed for its gentrifying effects. La Alianza Hispana (a social services agency) and the Riley Foundation (a charitable trust) proposed creating a new organization to revitalize the neighborhood without displacement. More than 200 community residents attended the first community meeting of the new Dudley Street Neighborhood Initiative on February 23, 1985. They objected to the composition of DNSI's governing board, which gave only four of the 23 governing positions to community residents. As a result, DSNI was restructured to provide a majority of board seats for community residents, divided evenly among the four major cultural groups in the area: African-American, Cape Verdean, Latino, and white. Now, about 600 community residents participate in DSNI's work.

The broader planning process began with small projects having concrete outcomes, which served to build community engagement and relationships with city officials and other stakeholders. The most effective of these projects was the "Don't Dump on Us" campaign, which focused on cleaning up and fencing vacant lots, and preventing the illegal dumping of trash in the area. The campaign was the result of resident surveys identifying immediate priorities, and it garnered the attention and support of then-Mayor Raymond Flynn.

Building on developing relationships and processes, as well as a growing set of information about Dudley Street Neighborhood conditions, DSNI turned to developing a master plan for the area. It sponsored community workshops, formed working groups and focus groups, held meetings between residents and subcommittees of the board, conducted door-to-door surveys, and conducted communitywide meetings. In addition to gathering more data about current conditions, DSNI received input on community needs and goals, as well as community residents' vision for the future of their neighborhood. DSNI produced a comprehensive 200-page master plan organized around housing, human services, and economic development. The plan contained 13 specific revitalization strategies, including:

- the development of 2,000 units of affordable housing;

- the creation of an urban village with a commons providing space for retail enterprises and recreation areas;

- the development of programs to stop the displacement of residents, to reorient the provision of social services, to address drug and crime problems, to provide child care, and to create business opportunities for entrepreneurs;

- the development of alternative financing methods;

- the mobilization of the neighborhood and the strengthening of racial, ethnic, and cultural identity and diversity;

- the creation of employment, training, and educational opportunities for residents; and

- the development of neighborhood-based businesses. (Quinones 1994, 756)

The BRA adopted DSNI's master plan as the official redevelopment plan for the area. The plan also supported DSNI's "Take a Stand, Own the Land" campaign, aimed at community acquisition and ownership of vacant parcels to provide the "critical mass" of sufficient contiguous land for redevelopment to support or to create markets for housing, businesses, retail enterprises, and services. DSNI created Dudley Neighbors Inc., an urban redevelopment corporation serving as a land trust to hold land for the community, with private long-term leases or private ownership of buildings but not land, to keep housing affordable and prevent gentrification. The city not only conveyed to Dudley Neighbors Inc. the land it owned in the neighborhood, but also granted it the power of eminent domain to acquire neglected and abandoned properties, the first example of a community-based nonprofit having the power of eminent domain. The ambitious, yet concrete and community-supported, master plan convinced city officials to grant the power. While the Riley Foundation's grants and a $2 million investment by the Ford Foundation provided initial seed money to support the planning and land acquisition efforts, DSNI has been able to assemble financing from many sources.

DSNI, working with area residents, the BRA, city entities, federal and state agencies, and other community groups, has sponsored about 400 new units of affordable housing, hundreds of thousands of square feet of new or renovated retail space, a town common, parks and playgrounds, community gardens, a multicultural festival, and improved delivery of social services. Property values have increased, while crime has declined. Although the work of building relationships, defining the process, assessing conditions, identifying goals, visioning the future, and implementing plans has occurred—and continues to occur—throughout an ongoing neighborhood planning process, all these elements played critical roles in this community-based planning success story.

Sources: Alves et al. 1995; Faber et al. 2002; Gilman 2005; Kob 2000; Quinones 1994; Simon 2002; Zielenbach 2003.

and alternatives. These goals, if identified independently of options and alternatives, could result in innovative generation of options and alternatives not initially conceived by the planners. In the example of the four vacant warehouses, a planning process that first identified the goal of diversifying the economic and job base of the community might lead the planning group to seek options and alternatives for four separate parcels with four separate types of business operations, instead of being directed towards an exciting but less helpful vision of a single use of the four parcels.

Stage 4: Visioning and projecting possible futures. The techniques that can be used to engage the participants in envisioning and designing possible futures for their community include design charrettes (i.e., a process by which a multidisciplinary team of professionals "works closely with stakeholders through a series of feedback loops, during which alternative concepts are developed, reviewed by stakeholders, and revised accordingly" (American Planning Association 2006, 57), scenario development, impact assessment, participatory land-use mapping, computer photo simulation, and visual survey techniques (Randolph 2004, 67–68). Visual presentation methods, small group and large group discussion of options, individual registration of preferences and consensus-building activities, and the use of maps and GIS software will enhance the quality of input received from community residents, as well as their understanding of the planning issues and their enthusiasm for the ultimate plan.

Stage 5: Making plans that can be implemented. Unfortunately, in the history of land-use planning, many plans have been developed but not adopted, or adopted but not implemented. Several techniques may reduce the risk of an ineffective plan. Elected and appointed officials can be involved early in the process. The media can be kept informed about the process and given copies of the proposed plan and the plan as adopted. The community residents can be involved in the implementation, especially through concrete, discrete projects, but also through advisory boards and enforcement monitoring. Planners can identify financial and other resources to implement the plan. At a minimum, though, the plan must be internally consistent, the local government must amend its zoning code and similar regulations to conform to the plan's content, and planners and officials need to evaluate specific land-use decisions to make sure they accord with the plan.

The contents of the plan may vary. The American Planning Association's Policy Guide on Neighborhood Collaborative Planning (available at www.planning.org/policyguides/neighborhood.htm. and summarized in Appendix B of this PAS Report) gives the following illustration:

Neighborhood plans and planning should address a wide range of issues, but should be tailored to meet their specific needs, for example:

a. A definition of neighborhood boundaries--a description of how they were derived and how they apply to municipal service areas

b. A directory of who is involved and who should be involved in the planning process

c. A vision statement

d. Overall objectives for each element of the vision statement

e. Physical plan of the neighborhood indicating proposed improvements to the neighborhood

f. Specific tasks and assignments

g. Design guidelines

h. Links to citywide objectives

i. A directory of resources

At a minimum, the plan must be internally consistent, the local government must amend its zoning code and similar regulations to conform to the plan's content, and planners and officials need to evaluate specific land-use decisions to make sure they accord with the plan.

j. Short-term implementation projects to build support and momentum

k. Statistics about the neighborhood, including population, employment, education, etc.

l. Maps showing neighborhood resources such as churches, libraries, parks, historic sites, neighborhood landmarks and characteristics such as demographics

m. An implementation chart

n. A date of adoption and date for the next review or update

o. Statement of acceptance by the municipality (American Planning Association 1998, Policy 24).

Winter Park, Florida, won the APA 2007 Innovation in Neighborhood Planning Award for its infill redevelopment plan for Hannibal Square. The plan preserved the character of this historically African-American area and added a diversity of affordable housing and many other community features with resident input. For example, this Habitat for Humanity bungalow respects local architecture, and the new Shady Park spray pool is a place of community gathering.

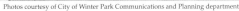
Photos courtesy of City of Winter Park Communications and Planning departments

At a minimum, neighborhood or district plans should contain planning principles, existing conditions, goals and needs, specific action items or desired outcomes, and implementation strategies. They should include narrative descriptions of these components, maps, tables and charts, and photographs or drawings comparing the current conditions with the planned outcomes. Those responsible for the plan need to employ a variety of means to disseminate the plans to interested parties. The Internet is a useful tool for making available large documents, but residents of low-income and minority communities may have limited access to the Internet, necessitating print copies and the use of libraries, community centers, and community-based organizations for disseminating the plans.

STARTING TO PLAN: DOING AN ENVIRONMENTAL JUSTICE AUDIT

Effective planning for environmental and land-use justice requires good information about environmental conditions in a locality's communities with a relatively high percentage of low-income people or people of color. A useful tool for gathering this information is an environmental justice audit. The audit provides a snapshot of demographic, historical and cultural, environmental, land-use, and economic facts about the selected area(s). It can be used for a variety of purposes, including:

1) identifying the environmental and land-use problems and planning needs of an area or the entire locality;

2) making the case for establishing a planning and regulatory program to seek environmental and land-use justice;

3) supporting specific land-use and planning decisions, especially the denial of development proposals in low-income or minority communities;

4) starting a neighborhood planning process or series of neighborhood planning processes in areas where environmental and land-use conditions are disproportionately burdensome; and

5) educating and involving public officials and the public about environmental and land-use injustices or problems.

In selecting the geographic area(s) for the environmental justice audit, planners may choose to perform the audit for the entire local jurisdiction. This choice has the advantages of treating all areas and neighborhoods equally, without regard to race or income, and gathering useful comparative data that can show how some areas bear a higher number of environmental harms or risks and have a lower number of public facilities and resources than do other areas.

Alternatively, planners may choose to select a particular area or set of particular areas (e.g., neighborhoods, districts) based on the following data and characteristics:

1. Socioeconomic demographics (race, ethnicity, poverty rate, income level, age, etc.) that indicate a heightened risk of exposure to environmental harms and risks

2. A history of zoning that shows a preponderance of intensive land uses or facilities with substantial environmental impacts

3. Environmental contamination or brownfields that require attention in context of broader planning efforts

4. Activism and interest of grassroots/neighborhood groups

5. Scheduling; namely, an area or subarea due for planning review as part of the jurisdiction's normal plan review process

The geographic scope of the area of study is likely to be a commonly recognized neighborhood, area, or corridor, as defined by both physical boundaries and sociocultural boundaries, or perhaps a preestablished planning area of the locality. The perceptions of residents, who often see the boundaries of a district or neighborhood differently than nonresidents, are

CHECKLIST FOR ENVIRONMENTAL JUSTICE AUDIT

General

1) May wish to do this audit across the entire locality

2) May wish to focus more exclusively on particular areas (neighborhoods, districts) from the areawide audit based on:
   - socioeconomic demographics (race, ethnicity, poverty rate, income level, age, etc.) that indicate risk of exposure to environmental harms and risks;
   - history of area being zoned for intensive land uses or the host of facilities with substantial environmental impacts;
   - locations of environmental contamination or brownfields that require attention in context of broader planning efforts;
   - activism and interest of grassroots/neighborhood groups;
   - part of the locality's periodic planning in areas or subareas of the locality (e.g., on a rotating basis or as simultaneous component parts of a jurisdictionwide comprehensive planning process); and
   - commonly recognized neighborhood, area, or corridor identity, as defined by both physical boundaries and sociocultural boundaries.

3) Use geographic information system (GIS) software

4) Be sure to include in audit immediately adjacent areas likely to have a substantial impact on the selected area (e.g., the major chemical plant across the street from a housing project on the edge of the study area)

Data to Gather

1) Demographic data (U.S. Census data):
   - Race and ethnicity
   - Income
   - Poverty level
   - Age
   - Type of household
   - Rates of homeownership

2) History and sociocultural features:
   - Area history, including land-use patterns, community identity, local residents, social and political movements, major events, and changes over time
   - Aesthetic and cultural assets/resources
   - Neighborhood groups
   - Major events
   - Historic structures
   - Social networks
   - Community strengths

- Environmental and land-use conditions
- Existing zoning designations
- Existing land uses (if different from zoning designations)
- Existing land-use plans for the area's future
- Superfund National Priority List sites
- Sites of hazardous-waste transportation, storage, and disposal facilities (TSDFs) under RCRA
- Five-year history of data from the Toxic Release Inventory (TRI)
- Available air-quality data
- Available water-quality data (both surface water and groundwater)
- Hydrologic patterns and flooding history (including sewer or stormwater overflow)
- Vacant or blighted sites
- Locations of schools
- Locations of parks
- Locations of civic centers and other public facilities
- Locations of sewage and water treatment facilities, power plants, power or gas distribution facilities, cellular towers, and similar facilities
- Conditions of streets, sewers, stormwater system, water distribution system, and distribution systems for electricity and natural gas
- Locations of airports, rail lines, ports/docks/marinas, mass transit routes, and other transportation facilities
- Locations of freeways, highways, and major arterial streets
- Emergency evacuation routes and emergency preparedness plans
- Locations of affordable housing stock (by type)
- Public health data on residents of area
- History of environmental and land-use problems or conflicts
- Economic conditions
- Major employers in area and number of area residents employed by these major employers (if data available)
- Employment/unemployment rates of area residents
- Income levels of residents
- Major economic producers and assets of area
- Community Reinvestment Act data on lending and investment in area
- Area residents' distance from work and their transportation options and choices
- Ranges and medians for rents and home values in area
- Education and skills levels of area residents
- Number and type of minority-owned businesses in area

The environmental justice audit should include immediately adjacent areas likely to have a substantial impact on the selected area.

also valuable. The environmental justice audit should include immediately adjacent areas likely to have a substantial impact on the selected area (e.g., the major chemical plant across the street from a housing project is on the edge of the study area). See the sidebar on the previous page for the potential types and sources of data that can specify areas for audits.

The written product of the environmental justice audit contains four components. An introduction identifies the audited area (including a map of the boundaries of the area in relationship to the locality as a whole), the purposes of the audit, the staff who prepared the audit, the range of dates of the audit research and preparation, and any other pertinent identifying information. A narrative summary of the information revealed by the audit follows. This summary both synthesizes individual sets of data and describes the socioeconomic, cultural, environmental, and land-use conditions of the audit area in an easy-to-read narrative. It calls the reader's attention to particular problems and inequitable distributions of harms, risks, benefits, and opportunities experienced by people in the audit area, as well as identifying the qualities, strengths, assets, and resources of the audit area. Then the audit data itself is presented in tables, lists, graphs, charts, and the like. Finally, a set of maps shows the geographic distribution of particular conditions, facilities, and land uses. The audit should make use of GIS software to map the relationships between data and location of the conditions or features that correlate to the data.

Ideally, the city or responsible organization will distribute the audit to a wide range of audiences, including audit-area residents and neighborhood groups, elected officials, planning commissioners and members of other relevant boards and commissions, the local government's chief administrative officer and other staff (including heads of relevant departments), and the media. Copies should also be available to the public at large, economic and civic groups (e.g., the chamber of commerce), businesses and industries in the audit area, and developers and/or property owners contemplating land-use changes in the audit area.

SPECIFIC ENVIRONMENTAL JUSTICE PLANNING ISSUES

Planning for environmental justice will likely examine specific issues that require special attention. Each of these issues is complex, and I do not attempt to provide comprehensive guidance on these issues in this publication. Rather, I briefly address them here because they are components of equitable land-use planning and regulation. Citations within the discussion will lead to further information. Please see the list of references in this report for full citations.

Transportation Planning

Resources on environmental justice and transportation planning are voluminous and growing (e.g., Bullard and Johnson 1997; Sanchez and Wolf 2005; Jakowitsch 2002). The issues largely break down into two categories: 1) disparities in access to transportation options, and 2) the location of freeways and heavily traveled roads, railways, and other transportation corridors.

In the first case, public transportation may be lacking altogether or may be primarily aimed at transporting suburban commuters to and from the central city, leaving central-city residents with limited service or no service to specific areas where low-income and minority people live, work, study, shop, play, or receive medical care. When examining access to transportation options for central-city residents, however, one must also examine street patterns, freeway access, sidewalks, maintenance of streets and sidewalks, pedestrian bridges, crosswalks, bike paths and lanes, and other similar issues of circulation infrastructure in and around low-income and minority communities.

Asthma Hospitalization Rates for Children 0-4 Years Old by Manhattan ZIP Codes (Rate Per 10,000)

- 0–55
- 64–92
- 96–142
- 146–234
- 257–505

Average Pediatric Asthma Rates (Per 10,000 Children):
- Northern Manhattan: 245
- New York City: 177
- Manhattan: 173

Source: NYC Department of Health SPARCS 2000 data on Asthma Admission Rates for children ages 0 to 4 years

Figure 3-2. *Asthma Hospitalization Rates by ZIP Code; Children Aged 0–4, Manhattan, 2000*

NORTHERN MANHATTAN FACILITIES

Map ID	Facility Name
1	Kingsbridge MTA Bus Depot
2*	DOS Garbage Truck Deport (two large depots; one services residents of the Upper East Side)
3	MTA Train Yards
4	Department of Transportation/Division of Highways Diesel Truck Depot
5	George Washington Bridge Port Authority Bus Terminal
6	North River Sewage Treatment Plant/ Riverbank State Park
7	135th Street Marine Waste Transfer Station
8	Manattanville MYA Bus Depot
9	Amersterdam MYA Bus Depot
10	Mother Clara Hale MTA Bus Depot (scheduled to expand)
11	DOS Garbage Truck Depot
12	126th Street MTA Bus Depot
13	Wards Island Sewage Treatment Plant
14	100th Street Bus Depot (currently expanding)
15	DOS Garbage Truck Parking Lot (outdoor parking lot)

NORTHERN MANHATTAN FACILITIES

Map ID	Facility Name
16	91st Street Marine Waste Transfer Station
17	59th Street Marine Waste Transfer Station
18	41st Street MTA Bus Depot
19	42nd Street Port Authority Bus Terminal
20	Hudson MTA Bus Depot (scheduled to close)

- ● MTA Bus Depot
- ▲ DOT Diesel Truck Depot
- ■ Marine Waste Transfer Station
- ⬟ Port Authority Bus Terminal
- ⬡ Department of Sanitation Facility
- ■ Sewage Treatment Plant
- ■ Train Yards
- /\/ Major Highways
- ᨀ 96th Street Demar

In the case of the siting of transportation infrastructure, the prevalence of highways in, through, and around low-income neighborhoods has resulted in heightened exposure to air pollutants and disruptive traffic flows for residents. Air quality "hot spots" often exist in neighborhoods close to freeways, and inner-city children experience disproportionate incidence of asthma (see Figure 3-2). Moreover, sprawl-producing land-use policies contribute to vehicular traffic, which in turn contributes to air pollution. Often, vehicle transfer, storage, or maintenance facilities are located in low-income and minority neighborhoods, further increasing the residents' exposure to air pollutants, as well as chemicals used at those facilities.

A major environmental justice issue is the particular vulnerability of low-income and minority communities to natural disasters and other emergencies, as well as the federal, state, and local governments' capacity and commitment to respond quickly, effectively, and fairly to the needs of low-income and minority communities in emergencies and disasters.

Impervious Cover

The amount of impervious cover, as well as land-use practices that generate pollutants, contribute substantially to urban stormwater runoff that floods urban areas and degrades the quality of urban waters and watersheds (Arnold 2005). Degraded and ignored urban watersheds, the "concrete jungle" environment of many inner-city neighborhoods, and the undersupply of parks, green space, and natural environments contribute to inner-city residents' stress, exposure to health risks, and lack of opportunity to connect to nature. Good planning for environmental justice addresses the amount of impervious cover in low-income and minority areas, the use of natural features in urban design, and the restoration and conservation of natural areas in the urban environment. In addition, it plans for management of urban runoff and rejects the false trade-off between developing greenfield watersheds in suburban and rural areas and further developing already developed and degraded watersheds in urban areas. A relatively new approach to handling the problems caused by run-off and pollution is known as "green infrastructure," which promotes the use of natural means to manage both stormwater and wastewater (Randolph 2004; Daniels and Daniels 2003; France 2002; Jeer et al. 1997. U.S. Environmental Protection Agency 2005; U.S. Environmental Protection Agency 2006).

Emergency Planning

As the plight of New Orleans' residents in Hurricane Katrina showed, a major environmental justice issue is the particular vulnerability of low-income and minority communities to natural disasters and other emergencies, as well as the federal, state, and local governments' capacity and commitment to respond quickly, effectively, and fairly to the needs of low-income and minority communities in emergencies and disasters (Farber and Chen 2006; Dreier 2006; Cutter et al. 2003; Peguero 2006; American Association of State Highway and Transportation Officials Center for Environmental Excellence 2006.). One policy report summarizes the situation as follows:

> It is society's most vulnerable who were "left behind" by government efforts to assess, to plan for, and to respond to a storm of Katrina's magnitude. And this was predictably so. . . . Twenty-eight percent of people in New Orleans live in poverty. Of these, 84 percent are African-American. Twenty-three percent of people five years and older living in New Orleans are disabled. An estimated 15,000 to 17,000 men, women, and children in New Orleans are homeless. The lowest lying areas of New Orleans tend to be populated by those without economic or political resources. The city's Lower Ninth Ward, for example, which was especially hard hit and completely inundated by water, is among its poorest and lowest lying areas. Ninety-eight percent of its residents are African-American. As Craig E. Colten, a geologist at Louisiana State University . . . explains: "[I]n New Orleans, water flows away from money. Those with resources who control where drainage goes have always chosen to live on the high ground. So the people in the low areas were the hardest hit." (Center for Progressive Regulation 2005, 34–35).

A much-neglected aspect of comprehensive planning, as well as area-specific planning, is the development of plans for emergency preparedness and response, as well as the creation of specific implementation tasks with timetables and mechanisms for ensuring tasks are completed. Even when such planning occurs, it may fail to consider worst-case scenarios and information about the particular challenges of low-income and minority communities, such as lack of cars, limited access to money, and physical conditions that might impair mobility. Emergency planning may also underestimate the roles of environmental conditions and land-use patterns, such as the locations of facilities using or producing toxic substances, either as potential sources of disasters or as factors that compound the harms to low-income and minority residents. These types of neglect in emergency planning might be the result of limited resources and the nonimmediacy of emergency or disaster conditions, but it ultimately comes at enormous human and institutional costs, as the Katrina disaster demonstrated (Comfort 2006).

Housing

A major land-use issue in low-income and minority communities is the supply and quality of affordable housing (Barnett 2003, 63-75; Russell 1996; Blackwell 2001). Planning and regulatory efforts that address environmental conditions should be coordinated with efforts to enhance the supply of affordable housing in those communities and to improve the quality of existing housing. One issue is where to locate new housing. Environmental justice planning calls for buffers between housing and industrial and other intensive land uses (including freeways) and for segregating those uses. Unfortunately, feasible sites for affordable and multifamily housing development may exist in close proximity to inappropriate land uses (e.g., remnant parcels from state and federal highway entities, adjacent to major freeways). These situations call for greater creativity, proactivity, and long-term planning by planners and stakeholders to avoid difficult trade-offs between the health and environmental conditions of affordable housing residents and the availability of affordable housing options. A similar issue is whether brownfields are appropriate sites for housing or other needed projects, like new schools and parks. This problem requires careful attention to the degree of knowledge and certainty about the types and levels of contamination on the property in question, the degree to which the property will be cleaned or remediated (including political commitments, private property owner commitments, and commitments of available financial resources and technical expertise), and the mechanisms for ensuring that consumers of affordable housing are not disproportionately bearing increased risks of exposure to hazardous and toxic substances due to brownfields redevelopment and housing policies. As this PAS Report mentions several times, the pressures to make economic use of land and the project proponents' inherently human tendencies to optimistically underestimate risk call for the use of risk-avoidance principles in planning for affordable housing.

Planning and regulatory efforts that address environmental conditions should be coordinated with efforts to enhance the supply of affordable housing in those communities and to improve the quality of existing housing.

The Difficulties of Redirecting Already Developed Areas

Environmental justice planning has the inherent challenge of planning for the future in the shadow of past planning injustices and existing, less-than-desirable land-use patterns. On one hand, aggressive government actions to remake an area face numerous potential problems: gentrification, displacement of current residents, and destruction of community identity; private property rights challenges, loss of businesses that contribute to the economy, and political backlash; tremendous investment of financial resources, staff time, and policy focus; and the risk that imposition of a plan will fail to match social and economic forces and changes that land-use policy cannot control. On the other hand, inaction or reliance solely on nongovernmental forces to change existing land uses will likely fail to remedy environmental injustice, solve current land-use problems, or achieve the goals and visions that community residents and public officials have for the area(s) in question. The use of regulatory tools like zoning changes, the application of environmental justice principles in land-use permit decisions, government investment in infrastructure that enhances the quality of the area(s), and persistent attention to implementing and achieving equitable plans will be necessary to effectuate change in already developed areas while sustaining existing community identity and assets.

CHAPTER 4

Regulatory Tools

Planning for environmental justice is not sufficient by itself. Incorporation of environmental justice principles into plans will have very little impact if government officials do not incorporate those principles and plans into the zoning code and map. Zoning that does not implement adopted plans for healthy, safe, and vibrant low-income neighborhoods of color may be illegal under the consistency doctrine, which requires that all zoning be in accordance with a comprehensive plan.

A well-grounded strategy for zoning or rezoning areas in which low-income and minority people are exposed to environmental harms and risks starts with three tools: 1) An environmental justice audit; 2) A zoning assessment; 3) Neighborhood-based planning.

Moreover, the exposure of low-income and minority communities to environmental harms and risks typically accompanies existing zoning classifications that permit intensive uses in those communities. Because people of color and the poor live near and among a higher proportion of industrial and commercial uses than do white, high-income people, an appropriate land-use planning response would be for cities, counties, and other land-use regulatory authorities to change the permitted uses in those areas to correspond more closely to the residents' desired neighborhood environment, as well as their health and safety needs.

AN ENVIRONMENTAL JUSTICE ZONING STRATEGY: GETTING STARTED

A well-grounded strategy for zoning or rezoning areas in which low-income and minority people are exposed to environmental harms and risks starts with three tools. The first is an *environmental justice audit* of each of the areas or neighborhoods likely to need new planning and changes to zoning to achieve environmental justice. The environmental justice audit is described in detail in Chapter 3 of this PAS Report. The environmental justice audit is a necessary tool not only for planning but also for rezoning.

The second tool is a *zoning assessment*, which is a systematic analysis of all of the permitted and conditionally permitted land uses in varying proximities to residential areas and other areas in which local residents will be exposed to environmental impacts, such as schools and day care facilities, parks and recreation areas, hospitals and clinics, and so forth.

The zoning assessment goes beyond the environmental justice audit. The environmental justice audit should provide information about the zoning designations of parcels in the area, as well as existing actual land uses. A zoning assessment, on the other hand, focuses on the specific types of uses allowed under the zoning designations of the area's various parcels. It builds on the zoning map assessment and parcel survey techniques by adding a zoning text analysis of all permissible or conditionally permissible land uses that could occupy parcels in close proximity to local residents.

The zoning assessment is used in evaluating whether text amendments, map amendments, or both are needed to protect low-income people of color from land uses with particular risks of harm or substantial adverse impacts. For example, an environmental justice audit may reveal that the C-2 zoning classification (i.e., a district allowing certain commercial uses by right) applies to 25 different parcels of land located within 100 yards of single- and multi-family residential land uses in a particular geographic area with an income level substantially below the area median income and a population of minority residents that substantially exceeds the median percentage of minority representation in the area. The zoning assessment might further reveal that C-2 zoning currently allows not only for retail, professional office, restaurant, and medical clinic uses, among others, but also allows for such potentially intensive land uses as machine repair and reassembly facilities, recycling transfer stations, beverage bottling and distribution plants, and electric power stations. Planners and community residents would use this information to assess whether the potentially intensive uses allowed in the C-2 zone are appropriately located within 100 yards of residences. This assessment might lead to recommendations to eliminate some of the permitted uses in C-2 zones, require conditional use permits for some of the uses currently permitted by right, or allow some of the uses, provided they are not within 100 yards of any residence.

The third tool is *neighborhood-based planning*, which should precede a rezoning strategy. Neighborhood-based planning is described in detail in Chapter 3 of this PAS Report. The plans, policies, and information resulting from a community-based planning process for a specific area or neighborhood may call for new zoning designations or changes to the uses allowed in particular zoning districts. Plans identify not only land-use problems to be solved, but also land-use opportunities to be pursued. A comprehensive rezoning process incorporates zoning changes that comprehensively address existing land-use problems and future land-use opportunities in low-income and minority neighborhoods.

ZONING AMENDMENTS

Strategies to make zoning more equitable are in essence strategies about changes to existing zoning. Zoning for environmental justice is a bit of a misnomer because officials are, in reality, rezoning for environmental justice. Indeed, existing zoning patterns and existing land-use patterns are sources of environmental injustice, and pose special challenges to achieving more equitable and sustainable patterns of land use. Despite these challenges, government officials and planners may want to consider text amendments, map amendments, or a comprehensive revision of both the text and the map.

Zoning Text Amendments

Local governments commonly use zoning text amendments to remove intensive uses from use districts in which those intensive uses are inappropriate in their view, without ever changing the district designation of any particular parcel. For example, a city council or county commission might amend the zoning code expressly to prohibit ready-mix concrete plants in I-2 (heavy industrial) districts (*Rockville Fuel & Feed Company v. City of Gaithersburg* 1972) or to change quarrying and extractive-type activities from "of right" uses in agricultural districts to conditional uses (*County Commissioners v. Arundel Corporation* 1990). In each of these cases, the designations of districts on the map did not change, but what was allowed in those districts changed through amendments to the permitted, conditional, and excluded uses that applied to all parcels bearing those designations. In addition, text amendments might have jurisdictionwide (i.e., multidistrict) applicability, as in the case of removing recycling operations from permitted uses in solid waste floating zones (*Free State Recycling System v. Board of County Commissioners* 1994) or classifying all airports, both commercial and noncommercial, as conditional uses in any district (*Von Lusch v. Board of County Commissioners* 1975).

Local governments might use zoning text changes to address the current or potential land-use problems of particular neighborhoods. For example, local officials might address a neighborhood with a checkerboard pattern of commercial uses by prohibiting electroplating businesses, solid waste incinerators, and machine shops in commercial zones. No parcel would lose its commercial use designation, but the range of permissible uses for commercial parcels would shrink. Similarly, a city might change a permitted "of right" use (e.g., metal foundries in industrial districts) to a conditional use, so that anyone seeking the use would have to obtain a conditional use permit and submit to certain conditions designed to protect the neighborhood.

Zoning text amendments have some legal advantages over zoning map amendments. Because text amendments are generally applicable and thus often deemed "comprehensive" in nature, they receive greater deference as legislative acts and are presumed valid (*Von Lusch v. Board of County Commissioners* 1975, 742; *Layne v. Zoning Board of Adjustment* 1983, 1,089). Even after a landowner has received a special exception that allows use of the property for an intensive

---

ZONING AND ENVIRONMENTAL JUSTICE: SOME QUESTIONS TO ASK

1) Are zoning patterns consistent with the environmental justice principles in the community's plans (implementation)?

2) Is your zoning compatible with environmental justice principles?

3) Have you examined your zoning patterns in light of environmental justice audits?

4) After performing steps 1 through 3, Do you need to amend the zoning code? Should these be text amendments, map amendments, or both?

5) When rezoning existing uses, have you considered the issues related to downzoning, spot zoning, and the creation of nonconforming uses?

6) Local governments will be most successful in defending the validity of rezoning commercial and industrial properties in low-income minority neighborhoods from more-intensive to less-intensive uses if they follow four guiding principles. Have you done or are you prepared to do the following?

  a. Rezone before controversial specific land-use proposals arise.

  b. Carefully document the incompatibility of existing high-intensity use designations and their impact or potential impact on the health and safety of local residents, as well as community character.

  c. Seek rezoning for all neighboring parcels with similar use designations and similar impacts (i.e., do not leave a landowner the argument that only his or her property has been downzoned while neighboring parcels remain zoned for more intensive uses).

  d. Avoid such intense downzoning that a landowner suffers a severe diminution in the property's value; leave the owner some economically viable use (e.g., downzone from an industrial use to a commercial use, instead of all the way to a single-family residential use).

7) Are you prepared to use both advanced techniques (see text) and permitting processes (see text) when needed?

Planning officials might address inappropriately intensive land uses and zoning designations in low-income and minority neighborhoods by amending the zoning map to change more-intensive use designations on parcels in these neighborhoods to less-intensive use designations.

land-use (e.g., a concrete batching plant in a heavy industrial zone), a city may prevent the use by amending the zoning code to prohibit the use altogether in heavy industrial zones; the landowner has no vested right in the continuation of any existing zoning (*Rockville Fuel & Feed Company v. City of Gaithersburg* 1972, 675–77). Text amendments do not address whether particular uses are appropriate on particular parcels singled out for attention, but instead are generally applicable determinations that certain uses are always incompatible with the other uses in a zoning classification or always need the oversight that accompanies conditional use permits. Thus, they avoid the potential problems associated with "spot zoning" (i.e., zoning a small area of land differently than surrounding land; see discussion below) and "downzoning" (i.e., changing more-intensive use designations to less-intensive use designations; see discussion below) that result from particularized treatment of individual parcels or small groups of parcels.

Please note, however, that local governing boards must follow required procedures and give affected parties proper notice and opportunity to be heard when adopting text amendments (*Free State Recycling System v. Board of County Commissioners* 1994, 806–08). In addition, changes to generally applicable zoning designations may arouse the opposition of many different affected landowners citywide, thus making them difficult to achieve politically. Furthermore, a text amendment may be too blunt a tool for excising intensive uses interspersed throughout low-income and minority neighborhoods. For example, a solid waste incinerator might be appropriate for most, perhaps even nearly all, heavy industrial zoning designations in a city. A text amendment to make it an impermissible use in industrial zones would not directly address the underlying environmental justice problem of industrial zoning in a residential area of color.

Zoning Map Amendments

Zoning map amendments change the zoning district designation for a particular parcel, tract of land, or set of parcels. Planning officials might address inappropriately intensive land uses and zoning designations in low-income and minority neighborhoods by amending the zoning map to change more-intensive use designations on parcels in these neighborhoods to less-intensive use designations. This technique is known as "downzoning." For example, a low-income minority neighborhood might contain several parcels zoned for heavy industrial use in close proximity to residences, schools, churches, health care facilities, and the like. The local government might rezone some or all of these parcels for less-intensive, yet economically viable, commercial uses.

Even though downzoning may change the land-use designations in low-income and minority communities to reduce threats to the residents' health, safety, quality of life, and sense of community, owners of downzoned parcels are likely to challenge the rezoning. A majority of courts will deem a rezoning a legislative act and give it a presumption of validity (Rohan 1998, Section 39.01[2], 39-4). Thus, the landowner will have to prove that the zoning amendment was "arbitrary, capricious or unreasonable and having no substantial relation to the public health, safety or general welfare" (Rohan 1998, Section 39.01[2], 39-4).

Courts in some states treat rezoning as an administrative or quasi-adjudicative act, making them subject to greater judicial scrutiny and requiring substantial evidence in the record to support the rezoning (*Neuberger v. City of Portland* 1979; *Cooper v. Board of County Commissioners* 1980; *Golden v. City of Overland Park* 1978). Courts in a few states require that governmental bodies support rezonings with evidence of either a substantial change in the character of the neighborhood where the rezoning occurred or a mistake in the existing zoning (*Wakefield v. Kraft* 1953; *Kimball v. Court of Common Council* 1961; *City of Biloxi v. Hilbert* 1992; *Seabrooke Partners v. City of Chesapeake* 1990; *Davis v. City of Albuquerque* 1982).

The "change or mistake" rule is problematic for addressing environmental injustice in low-income and minority neighborhoods. It creates strong inertia for existing zoning patterns, which are inequitably distributed and often harmful to low-income people and people of color. Attempts to change environmental conditions and land-use patterns by first changing zoning designations may be especially problematic. Local officials may not be able to support downzoning with evidence of changed conditions favoring less-intensive uses. Many low-income and minority neighborhoods contain the very industrial land uses and other LULUs that are inappropriate, yet pervasive. Moreover, the historic and ongoing presence of these land uses and poor environmental conditions contributes to the deterioration of the neighborhood, further undermining the argument that intensive land uses are now inconsistent with the area's emerging land-use patterns. Instead, local officials will have to argue that the initial zoning was a mistake based on invalid, or perhaps discriminatory, assumptions about the compatibility of industrial and commercial uses with nearby residential activities.

Even in the majority of states where rezoning is legally presumed valid, courts as a matter of practice scrutinize downzoning carefully. A landowner may contend that the rezoning is impermissible spot zoning or, more precisely, spot zoning in the reverse. Spot zoning results in a small area of land being zoned differently than the surrounding land, while spot zoning in the reverse entails zoning land more restrictively than the surrounding parcels. Spot zoning in the reverse, which is more relevant to the environmental justice goal of downzoning intensive uses in mixed use areas, is often struck down as arbitrary and capricious, an unjust discrimination against the downzoned parcel because surrounding parcels are not subject to the same treatment

A zoning text amendment may be too blunt a tool to use where intensive uses are interspersed in a neighborhood.

© iStockphoto.com / Matt Baker

(Reynolds 1995). Local officials must also take care to ensure that downzoning accords with, and does not facially conflict with, the comprehensive plan.

When a local government downzones property permissible under the prior zoning but now impermissible under the new zoning, the "nonconforming use" doctrine entitles the property owner to continue the existing use for at least a "reasonable" period of time. This doctrine protects the rights of private property owners to a return on their investment in constructing, developing, and/or commencing land uses under the laws that allowed the use at the time the use began. It prevents a local government, when it makes a zoning change, from demanding the immediate discontinuance of a use lawful at the time of the zoning change, unless the use is a public nuisance (*Livingston Rock & Gravel Company v. County of Los Angeles* 1954; *Oswalt v. County of Ramsey* 1985; *Dugas v. Town of Conway* 1984; *Bachman v. Zoning Hearing Board* 1985). The government,

Environmental justice land-use strategies might not effectively force changes in current actual land-use patterns, but instead would do so over time, as nonconforming uses cease to exist or are required to terminate at the end of an amortization period.

under the nonconforming use doctrine, may require the nonconforming use to cease after a reasonable "amortization" period, designed to balance the public interest in landowner conformance with the zoning laws against private property rights, particularly in the opportunity to obtain a reasonable return on the landowner's investment (*Standard Oil Co. v. City of Tallahassee* 1950; *City of Los Angeles v. Gage* 1954; *Harbison v. City of Buffalo* 1958).

An owner of a nonconforming use, in general, cannot change, extend, enlarge, or structurally alter the use, and will lose the right to the nonconformity if he or she abandons or discontinues the use or if the structures are totally destroyed (Rohan 1998, Section 41.03(1), 41–59). Therefore, environmental justice land-use strategies might not effectively force changes in current actual land-use patterns, but instead would do so over time, as nonconforming uses cease to exist or are required to terminate at the end of an amortization period. Local officials should monitor nonconforming land uses and develop an amortization and phase-out plan for them. As one expert states, cities "may wish to survey their nonconforming uses and determine whether any of them pose such health and environmental problems that they should be targeted for closures, either immediately as public nuisances or later through an amortization process" (Gerrard 2001, 148).

Furthermore, owners of downzoned property who suffer economic loss to accommodate neighborhood residents' opposition to their uses of their property may claim that the local land-use authority has unreasonably exercised its police power and has taken private property without just compensation. For example, downzoning a parcel from commercial to residential use was unconstitutional when it resulted in a 92 percent diminution in the parcel's value and nearby residences could be protected from the impacts of the business use of the land by an existing buffer area (*Grimpel Associates v. Cohalan* 1977). When the land-use planning authority can present sufficient evidence that downzoning is necessary to protect local residential neighborhoods, however, the downzoning is likely to be upheld (*McGowan v. Cohalan* 1977; *Moviematic Industries v. Board of County Commissioners* 1977). For a general treatment of takings law, see Chapter 8 of this PAS Report.

Low-income and minority neighborhoods are in something of a Catch-22. On one hand, zoning designations often reflect existing uses, which in the case of low-income and minority neighborhoods are often a set of mixed, intrusive, intensive, and even expulsive uses. Environmental justice advocates want to change these zoning patterns. Environmental injustice often affects older neighborhoods, however, and as Ellickson and Tarlock (1981, 59) observe, "[a]lthough all use designations are potentially amendable, those in established neighborhoods are the least likely to be open for negotiation." Amendments to the zoning code and zoning map are means of redefining acceptable land uses, at least for the future, but they will be judged by their compatibility with surrounding uses and the character of the neighborhood, which often reflect the very uses that grassroots groups are trying to change. Objectionable uses may be deemed compatible with nearby, similarly intensive uses. In addition, landowners who are accustomed to the intensive characterization of their parcels and the neighborhood are likely to resist change.

Local governments will be most successful in defending the validity of rezoning commercial and industrial properties in low-income minority neighborhoods from more-intensive to less-intensive uses if they follow four guiding principles:

1. Seek rezoning before controversial specific land-use proposals arise.

2. Carefully document the incompatibility of existing high-intensity use designations and their impact or potential impact on the health and safety of local residents, as well as community character.

3. Seek rezoning for all neighboring parcels with similar use designations and similar impacts (i.e., do not leave a landowner the argument that only his or her property has been downzoned while neighboring parcels remain zoned for more intensive uses).

4. Do not downzone so greatly that the landowner suffers a severe diminution in the property's value; leave the owner some economically viable use (e.g., downzone from an industrial use to a commercial use, instead of all the way to a single-family residential use).

Amendments to the Zoning Text, the Zoning Map, and the Comprehensive Plan

Perhaps the most successful strategy of all would be a comprehensive set of amendments to the zoning text, the zoning map, and the comprehensive plan. These combined text and map amendments often create new zoning designations and apply them to existing parcels, and they often receive judicial approval because of their comprehensive nature (e.g., *Jafay v. Board of County Commissioners* 1993, 898).

For example, if local planners and neighborhood residents were concerned that interspersed light industrial zoning might permit manufacturing activity with the presence and use of toxic chemicals, the emission of noise and dust, and the like, but did not object to warehouse uses (permitted uses in light industrial zones), they might consider four options. First, they could seek a zoning text amendment to delete manufacturing as a permitted use in light industrial zones. However, this change would seem to run contrary to the definition of light industrial activity as including at least some manufacturing and would likely develop opposition from manufacturers in other parts of the city whose property is zoned light industrial. Second, they could seek a zoning map amendment to downzone the area's light industrial property to commercial or residential. This would prevent manufacturing in the area, but it would also inefficiently and perhaps unjustly prevent owners of the downzoned parcel from using their land for warehouses, even though the residents have no objection to warehouses. Third, the planners and neighborhood residents could seek both a map and text amendment that would downzone the land to commercial but place warehouses among the permitted uses for commercial zones. However, warehouses might not be compatible with all other commercial uses, and residents and landowners in other parts of the city where there is commercial zoning might object to warehouses in their areas. Fourth, they could seek both a text amendment that creates a new "warehouse" zoning designation and a map amendment that rezones the light industrial properties to warehouse uses.

The creation of new districts accommodates the particular land-use compatibility needs of particular neighborhoods, such as low-income and minority communities that historically have suffered expulsive zoning and harmful land uses. It has the capacity to reflect changing social norms about what uses are deemed compatible and incompatible with other uses. It also increases the "supply" of zoning designations, perhaps avoiding inefficient and burdensome restrictions on land that result from attempts to avoid some uses in a particular classification's large number of permissible uses (which accompany a small set of use classifications). This method, however, risks proliferation of particularized use designations and piecemeal zoning. Overly specialized zoning designations could limit both the local community and the private landowner in options for the property's use if the conceived use is no longer viable or desired, or the property is to be sold. Nonetheless, communities may need to experiment with new zoning classifications in an attempt to achieve environmental justice.

The creation of new districts accommodates the particular land-use compatibility needs of particular neighborhoods, such as low-income and minority communities that historically have suffered expulsive zoning and harmful land uses.

Figure 4-1. More than 1,000 community residents and businesses have endorsed the Barrio Logan Vision, a community-driven plan that requires: zoning that separates industrial and sensitive uses; relocation of toxic industries away from residences; and affordable housing for current residents. The Barrio Logan Community Plan is 30 years old, the oldest in the City of San Diego. The Environmental Health Coalition of National City, California, has recently succeeded in getting the city to begin a community plan update to be completed in 2009.

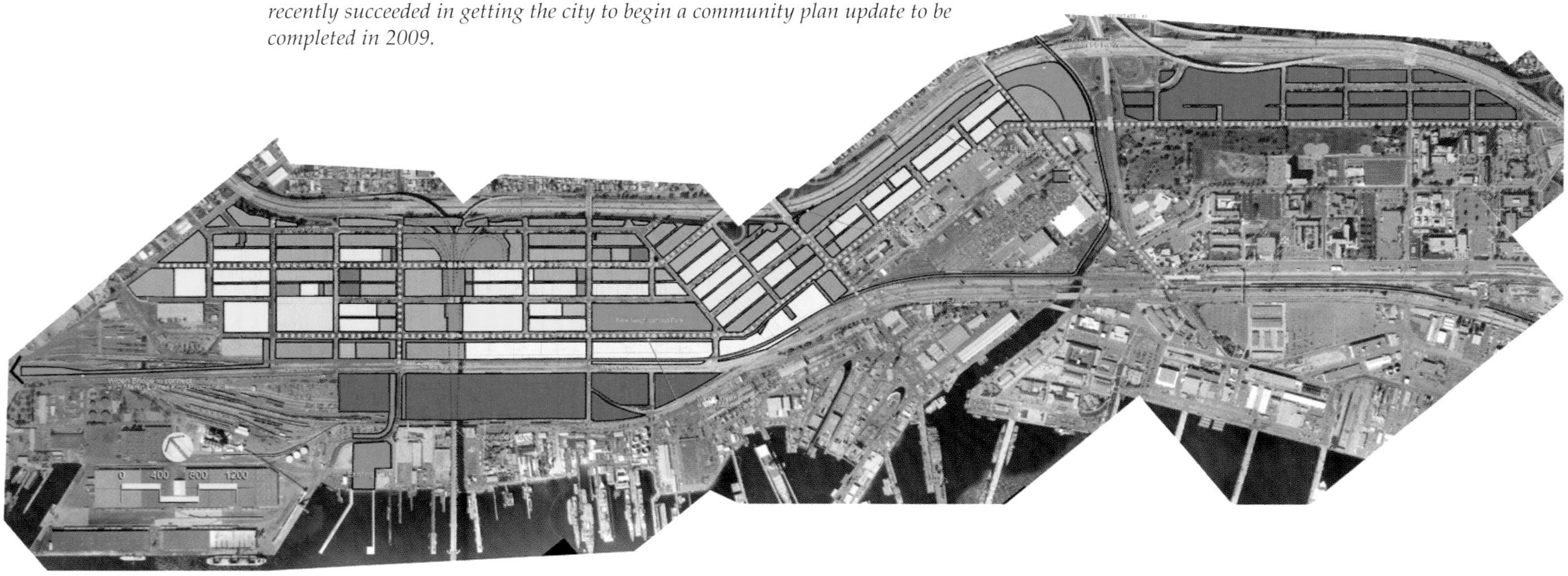

Environmental Health Coalition

Note: Existing housing, churches, and retail businesses will remain until voluntary change of land use or property ownership. The new use will then conform to the Vision.

COMMUNITY CENTER (4.62 acre)
FIRE STATION (0.20 acre)
HOSPITAL / MED CLINIC (2.33 acre)
INDUSTRIAL / MARITIME (71.57 acre)

PARKING (17.57 acre)
PARKS (51.56 acre)
RESIDENTIAL (53.63 acre)
RESIDENTIAL / RETAIL (22.61 acre)

RETAIL / OFFICE / COMMUNITY USE (43.36 acre)
SCHOOL (5.26 acre)
ENHANCE STREETS

In fact, a comprehensive zoning strategy for environmental justice likely will include a variety of components, including advanced zoning techniques (e.g., overlay zones and performance zoning), modifications to standards for granting, denying, and conditioning land-use permits (e.g., conditional use permits and variances), and perhaps even temporary moratoria to prevent further degradation of low-income and minority communities in the short-term while a comprehensive long-term strategy is being developed. These tools are discussed in the following sections.

ADVANCED ZONING TECHNIQUES

Tools for regulating local land uses for their environmental justice impacts extend well beyond traditional Euclidean zoning to a variety of advanced zoning techniques. These techniques offer land-use planners, local officials, and community residents many options for addressing current environmental injustices and preventing future inappropriate land uses in low-income communities of color.

Overlay Zones

An overlay zone is a means of imposing additional land-use requirements or restrictions on existing Euclidean zoning. "An outgrowth of Euclidean zoning, overlay zones in effect circumscribe an environmental area that is already subject to Euclidean regulation, and impose additional requirements thereon," writes Robert Blackwell (1989, 616). The additional requirements are laid over the existing zoning, so that the land in the overlay district is subject to the underlying traditional zoning requirements and the special requirements associated with the overlay district. Overlay zones have been used for a wide range of purposes, including prohibitions or limits on development where natural conditions (e.g., seismic hazard, hillside slope, or flood hazard) make it unsuitable; where there are aesthetic or historic features to be preserved; where sensitive and valuable environmental areas exist that could be harmed by excessive development; and where certain activities in the area (e.g., airplane flight patterns) make constraints on other activities necessary for safety or health.

Overlay zones can be used to impose a variety of specific requirements or restrictions on industrial and commercial land activities that occur in neighborhoods or areas inhabited by low-income people and people of color and that threaten the residents' health or the area's character and integrity. The specific additional requirements or restrictions imposed on the overlay zone will vary from locality to locality, depending on the concerns identified by local residents. The concept of the overlay zone allows these additional requirements or restrictions to be imposed only where they will help to protect and promote the health of the neighborhoods and the residents, not in other parts of the city where overlay zones might have no or little impact on residential areas. This narrow geographic tailoring of additional land-use regulations reinforces legal arguments that the regulations are designed to protect only those neighborhoods at risk of deterioration or environmental hazards without unnecessarily burdening land use in other areas. It also decreases the number of landowners citywide who might be affected and therefore might be opponents.

Overlay zones are often used in combination with other techniques, all of which are discussed in more detail below:

- *Buffer zoning* can be applied to areas where needed (e.g., where higher-intensity zoning or land uses border lower-intensity zoning or land uses). These could be termed "overlays of interface zones."

- *Performance standards* could be applied to specific areas where the environmental outputs of land uses are especially harmful, burdensome, or extensive, meaning all land uses in the area must meet specific performance standards, regardless of the underlying zoning designation.

- Overlay zones can be the means by which *neighborhood conservation districts* or similar area-protection districts are created.

- An overlay zone can be used to impose a *temporary moratorium* on specified land uses (e.g., all new industrial uses) in a specified area (e.g., a neighborhood or district with an overconcentration of such uses), until the city can develop a more comprehensive zoning scheme for the area.

- An overlay zone can require *conditional use permits* for certain land uses (e.g., yard waste storage yards) even if the underlying zoning (e.g., a heavy industrial zone, a light industrial zone, or an all-inclusive commercial zone) allows these uses by right.

- Finally, *transferable development rights* (TDRs), which authorizes landowners in certain designated areas to buy zoning rights (e.g., permitted building bulk) from landowners in other designated areas could be used to provide compensation to property owners whose land is downzoned or restricted by an overlay zoning ordinance in order to protect residents of a neighborhood affected by environmental injustice.

Land-use officials should be cautious about the inequitable use of buffer zones because, often, low-income, minority housing has served as a buffer between industrial uses and other uses.

The City of Austin made a particularly effective use of overlay zoning as part of an environmental justice land-use reform in East Austin, in combination with a limited moratorium, new conditional use permit requirements, neighborhood planning and rezoning, and enhanced neighborhood participation in land-use decision making. A case study of East Austin appears at the end of this chapter.

Buffer Zones

Buffer zones create a buffer or transition between a less-intensive use (e.g., single-family residential) and an adjacent or nearby more-intensive use (e.g., commercial or industrial) (Rohan 1998, Section 40.01[7], 40-38). The buffer zone between the two areas minimizes the impact of the more-intensive use on the less-intensive use. Low-income and minority residential neighborhoods need buffers to protect them from intensive industrial and commercial activity. Buffer zones can also include physical screening, landscaping, significant setbacks, open space, and even low-intensity commercial uses (e.g., offices, shops, churches, and medical care facilities). Local zoning codes might be amended to require specified types and distances of buffering between residential areas and an identified list of especially intensive or potentially harmful land uses. Planners, residents, and officials might identify land uses appropriate as buffers between residential and industrial areas, identify areas that need buffers, and link the areas with the uses on planning maps for rezoning and future development. Standards for screening and landscaping industrial and commercial facilities would also help to buffer visual and noise impacts on surrounding area residents.

Land-use officials should be cautious about the inequitable use of buffer zones because, often, low-income, minority housing has served as a buffer between industrial uses and other uses. In fact, the most frequent type of buffer between single-family residential areas and industrial or commercial areas is medium- or high-density residential uses (Rohan 1998, Section 40.01[7], 40-38). In the famous case of *Village of Arlington Heights v. Metropolitan Housing Development Corporation* (1977), in which the Supreme Court upheld the village's refusal to rezone land for low-income housing in an all-white Chicago suburb, the village's avowed purpose for its multifamily zoning designation was to serve as a buffer between single-family homes and commercial activities. Buffer zones are perhaps one of the major reasons why low-income and minority neighborhoods have so much industrial and commercial zoning: multifamily housing, where many low-income and minority people live, is purposefully placed near the industrial and commercial uses to create a buffer that protects high-income, white, single-family neighborhoods. Large numbers of poor and minority people have been placed near intensive uses, via zoning practices, because zoning and planning historically valued the single-family residence most of all, instead of the integrity and quality of all residential areas. Multifamily residences, schools, and parks and recreational areas where people will be exposed to the environmental outputs of nearby land uses should not be used as buffers. In addition, planners should use environmental impact analysis (discussed in Chapter 6) to evaluate whether distance, landscaping, screening, walls, and other such buffering techniques will adequately protect area residents from pervasive or migrating environmental conditions produced by intensive land uses.

Performance Zoning/Performance Standards

Performance zoning does not regulate land uses, but instead regulates the impacts of activities that occur on land (Kendig 1980; Acker 1991; Blackwell 1989; Rohan 1998, Section 40.01[7], 40-6). A performance zoning ordinance establishes standards for possible negative impacts on neighboring property (e.g., dust and smoke, noise, odor, vibration, toxic pollutants, runoff, glare and heat, and other nuisances—negative externalities). It prohibits any land use with impacts that exceed these levels which have been predetermined to be tolerable.

There are two ways of classifying performance standards. One is to distinguish between standards related to the effect of development density and design on natural resources—often associated with areas of new development—and standards related to the nuisance-like impacts of industrial activity(e.g., air, water, and soil pollution; noise; vibration; and odors)—often in established industrial areas. Another is to distinguish between what are known as "primitive" standards, which have only general definitions stemming from common law nuisance concepts (e.g., prohibitions on emission of "any offensive odor, dust, noxious gas, noise, vibration, smoke, heat or glare beyond the boundaries of the lot") and "precision" standards, which are developed from scientific data and reflected in quantifiable measurements (e.g., limits on permissible decibel levels in designated octave bands per second or designated frequencies in cycles per second).

All types of performance zoning ordinances merely supplement and do not replace traditional, use-based Euclidean zoning. Courts have largely upheld the validity of performance zoning standards both as reasonable means of protecting the public from nuisances and as sufficiently measurable according to a "reasonable person" nuisance standard (*State v. Zack* 1983; *Dube v. City of Chicago* 1955; *DeCoals Inc. v. Board of Zoning Appeals* 1981).

Performance zoning is essentially local environmental law. Except for the performance standards that prohibit all emissions, the standards permit some level of impact. Thus, performance standards by themselves do not address the problem of disproportionate industrial and commercial zoning in low-income and minority neighborhoods. Moreover, the permissible level of impact is based either on what is generally defined as "objectionable," which is vague and difficult to enforce, or on scientific calculations of risk. In either case, the standards require legal or scientific expertise, regulatory oversight, and control of pollution through risk assessment, rather than preventing pollution-generating sources near people's homes. Given the resources needed to develop, implement, and enforce performance standards, they are not as certain to keep pollution out—given slippages in enforcement and the potential for either careless or inadvertent emissions from heavy industrial activities—as prohibitions on industrial uses in these neighborhoods. Where industrial and intensive commercial uses cannot be entirely prevented or eliminated, performance standards offer a locally available tool for prohibiting those activities from polluting and disrupting neighborhoods. They are best used in coordination with strategies aimed at addressing land-use patterns in low-income and minority communities.

Neighborhood Conservation Districts

Neighborhood conservation districts are special zoning districts designed to protect older neighborhoods from the harmful or expulsive effects of mixed zoning with restrictions designed to prevent those impacts. According to a Texas appellate court in *Bell v. City of Waco* (1992, 214), "a neighborhood conservation district is an overlay district 'intended to encourage the continued vitality of older residential areas of the city, to promote the development of a variety of new housing of contemporary standards in existing neighborhoods, and to maintain a desirable residential environment and scale.'" The court upheld an ordinance enacted by Waco prohibiting the sale of automobiles on certain commercial property in neighborhood conservation districts, thus protecting these sensitive older areas from an arguably disruptive land use. Older neighborhoods of low-income and minority residents under environmental and development stress could be designated as neighborhood conservation districts, with specific regulations prohibiting or restricting uses that would have adverse impacts on the sustainability of these neighborhoods.

A similar, but more far-reaching concept is the "environmental preservation district." Collin and Collin (2005) call for "environmental preservation districts" in low-income communities of color as reparations for the environmental harms and land-use injustices these communities have suffered. They envision a district similar to a historic preservation district but focused on protecting the environmental quality, not historic character, of the community. "Like historic preservation districts, environmental preservation districts would not allow a property owner to demolish her property in order to put it to a more profitable use, would require her to restore the ecosystem if damaged, and would require her to go through a hearing before an environmental review board, similar to the hearings conducted by architectural review boards to address properties in historic districts" (p. 219). They would apply concepts of "carrying capacity" and the "precautionary principle" to prevent building out the district to full capacity with environmentally burdensome land uses (pp. 219–20). These districts could be applied to "the newly available urban lands—the sixty to seventy municipal landfills in urban areas that are closing, predominantly in East Coast cities," involving local residents in deciding the future of these properties and restoring them to environmentally sustainable and community-supporting land uses (p. 220).

Moratoria

A moratorium, usually enacted in the form of a zoning ordinance, either prohibits new development (sometimes of a particular type or in a particular area) for a period of time or imposes interim zoning restrictions that allow only limited development or require additional approvals (Selmi and Kushner 2004, 485). The purpose of a moratorium is to stop potentially burdensome or overintensive development (i.e., in type, in quantity, or both) from proceeding until the local jurisdiction can complete a systematic and well-informed planning process to make more permanent changes to local planning and zoning controls or complete the infrastructure necessary to support the development (Selmi and Kushner 2004, 485). These temporary

Collin and Collin (2005) call for "environmental preservation districts" in low-income communities of color as reparations for the environmental harms and land-use injustices these communities have suffered.

limits on new land uses and development allow the local government time to prepare well-conceived regulations and plans, without ill-conceived projects proceeding during the planning process and without government officials rushing to impose permanent regulations that either over- or under-regulate. In *Tahoe Sierra Preservation Council v. Tahoe Regional Planning Agency* (2002), the Supreme Court upheld the use of temporary moratoria as not constituting compensable takings of private property so long as the purposes are reasonable and there are reasonable time limits to absolute prohibitions on development.

In the service of environmentally just land-use regulation, a moratorium can be used to stop new industrial uses or other uses believed or known to be overconcentrated or overly burdensome in low-income and minority communities. The moratorium could be enacted as an interim overlay zone over the neighborhood or area in question, or as a jurisdictionwide temporary prohibition of all new development of a certain type within a certain distance of any residential use, school, house of worship, or medical care facility. Upon enacting the moratorium, though, local officials should promptly begin, and diligently pursue, a planning process to define appropriate uses and development standards for the area(s) or use(s) in question. The moratorium would eventually be replaced by permanent zoning code changes that would implement a well-planned and well-supported strategy to address the particular environmental justice and land-use issues that had arisen in the identified low-income and minority communities. See the case study about East Austin, Texas, above, for an example of the interim use of an overlay zone that required conditional use permits for certain industrial and commercial uses allowed by right in the East Austin area of Austin, Texas, pending a comprehensive planning and rezoning process for that area.

Although planning and traditional Euclidean zoning remain the foundational components of land governance and management in the U.S., much of "the action" lies in discretionary, often negotiated, land-use approvals, permits, and flexible techniques (Arnold 2005).

Transferable Development Rights

Transferable development rights (TDRs) are tools by which a local government "authorizes landowners in certain designated areas to buy zoning rights (e.g., permitted building bulk) from landowners in other designated areas. The (transferor) area is usually a historic district or other area the municipality desires to preserve. Landowners in that area are compensated for restrictions by being given TDRs, and the landowners in the transferee area bear the costs of the preservation program" (Ellickson and Been 2005, 92). Although typically used for historic districts and natural conservation areas, TDRs could be used as off-setting compensation to property owners whose land is downzoned or restricted in order to protect nearby residents. The receiving areas would need to be areas where residential communities would not be affected by the increase in intensity of use to which the transferee of TDRs would be entitled. In addition, officials should closely monitor any TDR program to ensure that low-income and minority areas are not TDR receiving areas so that they do not end up with more-intensive development in their communities than permitted by the underlying zoning.

DISCRETIONARY PERMITTING AND CONDITIONAL LAND-USE APPROVALS

For environmental justice planning principles and fair zoning practices to be effective, local decision makers must implement them when making specific decisions about proposed projects and land uses. Although planning and traditional Euclidean zoning remain the foundational components of land governance and management in the U.S., much of "the action" lies in discretionary, often negotiated, land-use approvals, permits, and flexible techniques (Arnold 2005).

Conditional uses are not a means of excluding potentially harmful activities from areas zoned for them because the zoning code lists them as permissible if they meet certain conditions, thus presuming general compatibility.

Opportunities for Achieving Environmental Justice Using Project-Specific Approvals

Project-specific discretionary land-use approvals offer several opportunities to integrate environmental justice into land-use decisions and policies. These opportunities include the following:

1. Project-specific analysis of environmental, land-use, and social impacts

2. Conditions of approval aimed at eliminating or reducing the adverse environmental impacts of the particular project, among which are project redesign, operational conditions, mitigation measures, and similar methods of defining or limiting the design and operation of the land-use in question

3. Public participation in the review and decision making about land uses that require government approvals

4. Problem-specific collaborative problem solving

5. Creation of incentives for clean, low-impact commercial and industrial uses, especially to the extent that some commercial and industrial zoning will remain in proximity to low-income and minority neighborhoods

6. Development of and mandates for "best management practices" for particular land uses to address impacts on surrounding neighborhoods

7. Permit renewals that require conformity to new standards (and parity between new facilities and existing facilities)

Local governments use a variety of discretionary land-use approvals into which environmental justice considerations might be integrated. They are:

1. conditional use permits (special exceptions, use permits, etc.);

2. variances;

3. subdivision approvals, planned unit developments (PUDs), and development agreements;

4. building, grading, and construction permits;

5. floating zones; and

6. environmental permits.

Conditional Use Permits

Most zoning identifies some uses that are permitted in the zone only if the landowner obtains a permit and meets the standards or conditions listed in the code for those uses (Rohan 1998, Section 44.01[1], p. 44-1 to 44-3). These uses are often compatible with other uses in the zone but are not necessarily compatible in every location or under every circumstance, or without certain limitations and conditions. The terms "special permits," "special exceptions," and "conditional use permits" are legally the same and are used interchangeably. Conditional uses are not a means of excluding potentially harmful activities from areas zoned for them because the zoning code lists them as permissible if they meet certain conditions, thus presuming general compatibility. Instead, conditional uses are a means of imposing certain restrictions on uses that could become nuisances or unduly burdensome on the surrounding area if left unchecked. They also allow for greater public scrutiny of some land-use proposals.

Conditional use zoning poses the risk that environmentally harmful land uses will be approved for low-income and minority neighborhoods if the residents' participation level or political strength is low and/or if decision makers do not consider the environmental justice impacts of these proposed uses when deciding whether to approve them, and if so, what conditions to impose on them. Environmental justice litigation has arisen in cases in which land uses like landfills and hazardous waste incinerators received conditional use permits from local governments (*East-Bibb Twiggs Neighborhood Association v. Macon Bibb Planning & Zoning Commission* 1989; *Security Environmental Systems, Inc. v. South Coast Air Quality Management District* 1991).

Conditional use zoning, however, also creates the potential for better control over potentially intensive land uses than by-right uses in industrial and commercial zoning districts, provided that the standards for and actual practices of deciding conditional use permits incorporate environmental justice considerations. These considerations are discussed below under standards for the grant or denial of land-use approvals, analysis of impacts, conditions of approval, participation, and the precautionary principle in light of inaccurate or incomplete information.

Variances

The broad discretion of zoning officials to grant variances from the requirements and restrictions of the zoning code poses the potential for harmful land uses (use variances), inappropriate structures and site developments (area variances), and inappropriate operations (other possible variances from regulations of land-use operations like hours, security, noise, etc.). Standards for granting variances like "unnecessary hardship" and "practical difficulties" are broad and open to interpretation by officials (Mandelker 2003, Sections 6.39-6.52, 6-44 to 6-6-60). Owners or developers of industrial facilities or other LULUs in low-income and minority neighborhoods may claim they cannot earn a reasonable return if they have to comply with zoning regulations. They may point to the presence of existing similar uses in the area that enjoy land-use privileges they do not and may argue that, given the existing mix of uses in the area, use of the land in less-intensive ways is not feasible or practical. Buffering and setback requirements are especially vulnerable to area variances and are often justified on the basis of the irregular shape of the lot or the location and frontage of the lot.

The inappropriate grant of variances, however, can be effectively constrained by standards that protect area residents. Local governments may adopt standards requiring that the variance "'not alter the essential character of the locality'" or have a substantial adverse impact on the neighborhood (Mandelker 2003, Section 6.47, 6-52). A comprehensive environmental justice strategy might include zoning code language governing decisions on variances that would authorize denial of the variance if granting it would impose a substantial risk of harm to the public health, safety, or character of nearby residential areas. The language might expressly authorize the consideration of cumulative impacts and overconcentration of nonconforming uses and structures in the area. In addition, area residents should be entitled to notice of hearings on variances, and these hearings should be held at times and in places convenient for the neighbors. Finally, conditions that minimize or mitigate the land use's adverse impacts can be imposed on the land use receiving the variance (Mandelker 2003, Section 6.51, 6-57 to 6-58).

Subdivision Approvals, Planned Unit Development (PUD) Approvals, and Development Agreements

Subdivision and PUD approvals will typically not address significant environmental justice concerns because they usually apply to residential development or mixed-use developments in locations other than low-income and minority neighborhoods. Nonetheless, it may be appropriate to consider environmental justice in two of the ways these approvals are used. The first is when the primary decision about development of a housing project in a low-income or minority neighborhood or one that specifically serves low-income people and people of color occurs at the subdivision or PUD approval stage. In these circumstances, the decision makers should give particular attention to the project's proximity to industrial and other intensive land uses, the environmental conditions of the surrounding area, and the environmental conditions of the property itself. The second is when an industrial development or a potentially intensive commercial development in or near a low-income or minority neighborhood is being developed as a PUD or requires the subdivision of land. In these cases, a PUD approval and/or subdivision approval would essentially be an approval of the project itself, necessitating consideration of impacts on the community.

Similar to PUDs, development agreements serve as project-specific zoning to a particular site or area, with the specific details of the agreed-upon development project becoming the terms and conditions of the applicable land-use regulation. These are negotiated approvals that guarantee the developer vested rights in the terms of the agreement (within some parameters), while allowing the local government to tailor the conditions and scope of its requirements and restrictions to the project's specific characteristics and context, as well as the locality's particular needs. Consideration of environmental justice principles in the negotiation and approval of development agreements requires attention to the project's impacts on low-income and minority communities and the participation of residents in impacted neighborhoods in the negotiation and approval process.

Building, Grading, and Construction Permits

Building, grading, and construction permits are ministerial permits in many jurisdictions and do not involve much opportunity for discretionary judgment or prevention of environmental injustice if the applicant meets all the requirements for the permits. However, they might be used to protect low-income communities and communities of color in two circumstances. One is when the locality uses any of these permits for discretionary decision making about the acceptability of the proposed development or use. Some jurisdictions might do so, and in this situation, the permit decision is much akin to a conditional use permit decision. The other is to ensure compliance with zoning requirements and conditions of approval in already-granted discretionary permits. If a permit (or the zoning code) requires that the landowner or developer install certain stormwater runoff detention basins and filters, submit a neighborhood-specific emergency response plan, and provide landscaping buffers at the edge of the property, for example, local officials usually can deny a ministerial permit if the applicant has failed to comply with these conditions.

Floating Zones

Floating zones are flexible zoning techniques that require particular scrutiny and monitoring by environmental justice groups to ensure that low-income communities and neighborhoods of color are not assigned harmful or burdensome floating uses. A floating zone is a land-use district created in the zoning code text but not yet designated on the zoning map (Callies et al. 1994, 69). The zoning authority identifies a need for a particular type of use but may not be able to identify where in the locality that use should be placed or zoned. Rather than be limited by the rigidity of traditional Euclidean zoning, the authority creates a district without any specific location(s) on the map, but with a set of standards for determining appropriate locations. The zone "floats" until a landowner seeks to have it applied to a property via a rezoning of the property. Thus, there is a "split" between the creation of the zone and the application of the zone to any specific areas. It gives the local authority flexibility in responding to local land-use needs.

Floating zones appear to be used most often for either industrial uses or high-density residential uses. For example, in *McQuail v. Shell Oil Co.* (1962), New Castle County, Delaware, applied an industrial floating zone to an undeveloped parcel that had been zoned residential, so that Shell Oil could build a refinery. Residents of low-income and minority neighborhoods may find that property zoned for nonintensive uses (e.g., residential) may be rezoned for industrial uses through the application of a floating zone at the request of the landowner. In order to ensure that floating zones are not used inappropriately to locate industrial uses in low-income and minority communities, local officials should establish criteria for the application of floating zones that require the consideration of environmental impacts and impacts on community character and integrity in determining the propriety of attaching the land-use to a particular location, the possible need for buffer zones between these uses and residential land uses, and the imposition of local environmental permits to control design and operation so as to minimize and mitigate harms.

Environmental Permits

Some governments have ordinances requiring local environmental permits for specific types of uses with specific types of environmental impacts. The use of environmental permits can be a highly effective way of evaluating and controlling the impacts of these types of uses on low-income and minority communities, if used properly. A local environmental justice strategy might consider the adoption of environmental permit requirements with specific environmental justice criteria included in the ordinance. However, environmental permits are not an adequate substitute for incorporating environmental justice considerations into land-use permitting because the scope of land uses having potentially adverse impacts on low-income and minority communities and the scope of those impacts are considerably broader than are usually the subject of environmental permit requirements.

STANDARDS FOR THE GRANT OR DENIAL OF DISCRETIONARY PERMITS

One critically important way to incorporate environmental justice principles into land-use decisions is to consider the impacts of proposed projects when granting or denying discretionary permits. The permitting process is designed to control the adverse impacts of proposed projects and land uses on the surrounding area and on the public.

Existing standards for granting or denying land-use permits likely give decision makers considerable discretion to deny a project that will have substantial adverse impacts on a low-income or minority community. Many

GRANTING, DENYING, AND USING DISCRETIONARY PERMITS: SOME CAVEATS AND SUGGESTIONS

- Consistency with environmental justice principles in plans is a reason to grant a permit (i.e., the permit implements the plan).

- Environmental harms and injustices are reasons for denial.

- Cumulative and synergistic impacts of a use on the area or surrounding uses are reasons for denial or imposing conditions (e.g., in "best management practices"); these conditions might include project redesign, mitigation, and specific changes related to expressed community concerns.

- Compliance with "best management practices" for particular uses, including buffers and performance standards, can address the impacts on surrounding neighborhoods and should improve chances of permit approval.

- The precautionary principle (i.e., regulators do not need scientific certainty that a particular course of action will have adverse consequences in order to take action to prevent the action if it poses a substantial risk of adverse consequences) should be employed if any doubt remains as to the completeness or accuracy of the information submitted with a permit application.

- Decision makers can give strength to permit approval by mandating a time limit on the permit, which would allow review at renewal times, ensuring both conformity to updated standards and parity with and for other uses.

- Use an environmental justice audit or environmental impact assessment to emphasize and to ensure the legitimacy of community concerns; such an analysis of impacts also provides a solid legal foundation for establishing permit conditions or for rejecting the permit.

- The permit process can ensure the early and meaningful involvement of area residents in an open and accessible decision-making forum. The process should promote full participation of community residents in hearings and discussions, as well as collaborative problem-solving and multiparty negotiation. (See Chapter 5 for more on participation.)

- Exactions and impact fees can be used to compensate residents of an area that is subject to a use requiring a discretionary permit, and these fees can be incorporated in the terms of the permit.

INDUSTRIAL REZONING, CUMULATIVE IMPACTS, AND BUFFER ZONES, NORTH DENVER NEIGHBORHOODS, DENVER, COLORADO

In the late 1980s and early 1990s, Denver undertook a citywide revision of its industrial zoning standards in response to concerns by environmental justice groups about the overconcentration of industrial uses in North Denver neighborhoods.

In October 1987, a coalition of grassroots groups, mixed-race but composed primarily of Hispanic residents of three neighborhoods (Elyria, Globeville, and Fwansea), formed an organization called "Neighbors for a Toxic Free Community." The group identified the archaic content of the industrial zoning code as one reason for the existence of so many locally unwanted land uses (LULUs) in their neighborhoods. The coalition, armed with hard data on the saturation of LULUs in low-income minority neighborhoods, was successful in obtaining support from the local housing authority, schools, and political leaders, including a state senator. In 1989 and 1990, the activists and city officials developed several amendments to the industrial zoning code, which the city council approved unanimously.

These amendments include requirements that:

- industrial uses be separated from residences by buffers;

- local residents be notified about and have an opportunity to comment on applications for industrial uses or hazardous materials storage; and

- the zoning administrator have the authority to deny a permit based solely on the area's undue saturation with uses that manufacture, use, or store hazardous materials.

In addition, an environmental review committee was established to review proposed land uses that involve hazardous materials; the committee can withhold a permit if it agrees unanimously to do so.

Despite limited enforcement, these amendments made a difference in at least one case. The Denver Board of Adjustment for Zoning Appeals reversed the zoning administrator's grant of a conditional use permit for Laidlaw Environmental Services to operate a solid waste transfer station in an I-2 zone. A neighborhood group, Park Hill for Safe Neighborhoods, with the help of the Sierra Club Legal Defense Fund and the Land and Water Fund, opposed the permit. The groups argued the permit should be denied because of an undue concentration of neighborhood uses having hazardous substances, not merely releasing hazardous wastes. The Denver Board of Adjustment agreed with their arguments, and a Colorado District Court affirmed the board's decision. The court deferred to the board's interpretation of the industrial zoning code's undue concentration provision as reasonable, within its authority, and supported by the evidence. ▪

Sources: Granado 1997; COPEEN 2000; Denver Board of Adjustment 1995; Laidlaw Environmental Services, Inc. v. Board of Adjustments 1996.

state zoning enabling statutes or local zoning codes authorize approval of discretionary or conditional permits if the proposed project is compatible with the surrounding area, is consistent with the public interest, or does not harm the public health, safety, morals, and welfare (Mandelker 2003, Sections 6.39-6.6.56, 6-44 to 6-67). If a landowner can show he or she has met the criteria for the permit in question, courts will consider the owner entitled to the permit and will reverse a local government's denial of the permit (*Zylka v. City of Crystal* 1969; *Bankoff v. Board of Adjustment* 1994). However, the burden of proof is on the applicant. If local decision makers find that the applicant has not met the criteria, based on substantial evidence in the record, they are not only entitled to deny the permit, but also are required to deny the permit. Thus, local governments arguably do not have the authority to approve discretionary permits for uses incompatible with the surrounding area, inconsistent with the public interest, or injurious to the public health, safety, morals, or welfare, depending on what the particular locality's ordinances provide as criteria.

Cumulative impact analysis, showing that a project will contribute impacts beyond a saturation point for the area, is a legitimate basis for a denial of a permit, despite the usual legal expectation that similarly situated land-use applications should be treated similarly (Rieser 1987; *Fawn Builders, Inc. v. Planning Board of Town of Lewisboro* 1996; *Buechel v. State Department of Ecology* 1994; *Jurgenson v. County Court for Union County* 1979). For example, the Denver Board of Adjustment reversed a grant of a conditional use permit for a solid waste transfer facility in an industrial (I-2) zone in a low-income neighborhood of color, in part because "[t]he area in which the station is to be located has an undue concentration of uses which manufacture, use, or store materials which create environmental hazards" (*Board of Adjustment for Zoning Appeals of the City and County of Denver* 1995).

Despite the broad authority many localities possess to consider environmental justice impacts in deciding on discretionary permit applications, a sound environmental justice policy would amend the permit criteria in zoning codes, to the extent allowed by state enabling statutes and local government charters, to include specific environmental justice considerations. An example might be as follows:

> *Environmental Justice.*
> To establish that the proposed land-use or project meets the criteria for the grant of [type of permit] under Section _____ of this Code concerning [compatibility with the surrounding area; injury to the public health, safety, morals, and welfare; consistency with the public interest; or similar wording that tracks the locality's criteria for granting or denying the particular permit], the applicant must establish:
>
> 1. the proposed land use or project does not pose substantial risk of harm to the health or safety of people who live, attend school, enjoy recreational facilities, or receive medical treatment or care within one mile from the site of the proposed land-use or project;

2. the applicant has adequately eliminated or minimized any pollution likely to be generated from the project, land use, or site, to the maximum extent practicable; and to the extent that the elimination or minimization of pollution generation is not practicable, the applicant has adequately mitigated its adverse impacts, provided however that no land use listed in Section _____ [a list of industrial and waste-related land uses especially intensive and potentially harmful, such as chrome-plating facilities] shall be permitted within _____ feet [select distance based on the need for buffering from the facilities on the list] of any property zoned or used for residential use, measured by the closest distance from the closest boundary of the parcel zoned or used for residential use to the closest point of any building, facility, or operational equipment used for any use listed in Section _____ [designed to allow the applicant or landowner to set aside a portion of a large lot for nonindustrial use (e.g., landscaping) as a buffer, instead of measuring to the lot line of the industrial property];

3. the proposed land use or project does not contribute to a disproportionate burden on any community with a relatively high percentage of low-income persons or persons of racial or ethnic minority groups, taking into consideration cumulative impacts and synergistic impacts of all land uses and environmental conditions affecting any such community;

4. the infrastructure in the area and the locality are adequate to support the proposed land use or project, in light of all existing and projected demands on the infrastructure from all sources;

5. the proposed land use or project will not constitute a nuisance; and

7. the proposed land use or project is consistent with any applicable comprehensive or area-specific plans, including any environmental justice principles contained in these plans.

Photos by Martha Arosemena

The lack of planning and the failure to regulate land-use patterns results in substandard and unhealthy conditions, such as in Texas colonias, which lack basic infrastructure.

In addition to environmental justice standards for the grant or denial of permits, local officials may want to consider developing and requiring a set of "best management practices" (BMPs) for particular land uses to address their impacts on surrounding neighborhoods.

In addition to environmental justice standards for the grant or denial of permits, local officials may want to consider developing and requiring a set of "best management practices" (BMPs) for particular land uses to address their impacts on surrounding neighborhoods. Buffers and performance standards would be especially appropriate. The particular BMPs might vary according to the particular land use in question, its potential impacts, its typical design and operations, and the history of its burden on low-income and minority communities. Therefore, BMPs for a dry cleaning facility may be different than BMPs for an automotive repair facility. Permits could be conditioned on compliance with the BMPs (see below), but a project proposed without meeting the BMP standards could be denied if the zoning code so provided.

Finally, local officials might want to consider time limits on the grant of discretionary approvals to potentially intensive land uses. This would allow the locality to use permit renewals to require the use to conform to operational standards as they emerge, incorporating new information about impacts and methods for controlling or mitigating those impacts. They would also allow for greater parity between facilities permitted under a time-limit-and-renewal system, which would impose new standards at permit renewal times, and future facilities that must comply with new standards.

Analysis of Impacts

In addition to specific permitting standards to address environmental justice concerns, local officials should give special attention to evidence that a proposed project or land use is likely to have an adverse or disproportionate impact on low-income or minority communities. This practice involves gathering information about the environmental and social impacts of proposed projects, analyzing the data, considering this information in relationship to the permit criteria and to the locality's environmental justice policies, and making findings that articulate the reasons for the grant or denial of the permit in light of the evidence. Local officials do not need irrefutable proof or analytically rigorous quantitative studies, but they do need more than just statements of community concern to establish a sound basis for denying a project likely to have adverse impacts on a vulnerable community. Further details about conducting environmental impact assessments that give attention to environmental justice concerns can be found in Chapter 6 of this PAS Report.

Conditions of Approval

Principles of environmental justice and equity in land-use decision making also require diligence in designing conditions of approval that minimize and mitigate adverse environmental impacts on low-income communities of color. Although the prevention of land uses likely to harm these communities should be the preferred approach, many land uses may be acceptable to, even welcomed by, community residents if they are appropriately conditioned. Environmental justice analysis can be built into project redesign requirements that result in the prevention or minimization of adverse environmental and social impacts. Environmental justice analysis can also be built into mitigation measures, which offset risks of adverse impacts. Specific conditions that address environmental justice concerns will vary with the type of land use, the particulars of the project proposal and design, existing conditions in the area, and the characteristics of the community where the project is proposed. Community concerns (expressed in various participatory venues), environmental impact assessments, and environmental justice audits are valuable sources of ideas about appropriate conditions of approval. In addition, the time-limit-and-renewal system of putting a time limit on permits and requiring owners to seek renewals offers the potential to reassess the conditions of approval and to develop new or modified conditions as information and circumstance warrant.

Participation

The process of discretionary land-use permitting is particularly vulnerable to underparticipation by low-income and minority area residents, and planners and officials may need, therefore, to make particular efforts to enhance participation, as discussed in Chapter 5 of this PAS Report. In particular, the locality may require project applicants to hold early, substantive meetings with area residents, even before filing for a permit. The Louisville Metropolitan government requires permit applicants to meet with residents of the area surrounding the proposed project and to document the meeting(s) prior to being allowed to file the permit application. The locality should also give special attention to ensuring that the process of discretionary land-use decision making is open and accessible to the public, including residents of the area affected, and that community residents have opportunities to participate fully and meaningfully in hearings on the permit(s). Finally, many land-use projects now result from negotiated or collaborative processes between regulators/planners and landowners/developers. Community residents should be involved in these processes as coequal participants. In addition, where particular conflicts or problems exist over a proposed project's potential impacts on the community, collaborative problem solving or multiparty negotiation processes must include grassroots community groups and neighborhood leaders.

One of the primary issues in evaluating a proposed land use or facility in a low-income or minority neighborhood involves inaccurate or incomplete information about the project's likely impacts on people, the neighborhood, the local community, and the environment.

Inaccurate or Incomplete Information and the Precautionary Principle

One of the primary issues in evaluating a proposed land use or facility in a low-income or minority neighborhood involves inaccurate or incomplete information about the project's likely impacts on people, the neighborhood, the local community, and the environment. Anyone who has been involved in assessing, reviewing, or deciding on land-use permit applications is well aware that approved land uses can ultimately have adverse impacts not intended or desired by government officials and planners due to any of seven informational constraints, described here:

1. *False information.* Unfortunately, some project applicants, or their consultants, present information they know or suspect to be inaccurate, hoping that no one will detect the errors.

2. *Omitted information.* Some project applicants, or their consultants, fail to provide relevant information that indicates potential adverse impacts of the project in question.

3. *Overly optimistic predictions and assessments.* Many project applicants, and their consultants, look at—and present—information about the project's impacts in a relatively positive light, in some cases underestimating the risk, magnitude, types, or scope of possible impacts.

4. *Acknowledged uncertainty.* Some types of desired or relevant information are not known, known only with low to moderate levels of certainty, or even not feasibly knowable. Many land-use decisions must be made with acknowledged uncertainty about their full impacts.

5. *Unanticipated consequences.* The very nature of this informational constraint is that no one expects the ultimate impacts to occur. We may lack knowledge, past experience, or analytical tools to predict the consequences, or conditions may change in unexpected ways. Nonetheless, we know that some land-use decisions will have unanticipated consequences, even if we cannot predict what they will be.

6. *Underused or misused information.* Project applicants, planning experts, and government officials may have access to relevant information, but may not use it effectively in making decisions. One reason simply involves the limits of human cognition. We are limited in our capacity to absorb, process, and use information in making judgments, and frequently other demands on our information-processing abilities can result in overlooking or misunderstanding potential impacts of proposed land uses.

7. *Inherent slippage and variation.* A frequent problem with land uses with adverse impacts on neighbors, the community, the environment, and/or the public infrastructure is that the project is not built or operated as represented and approved. Needs and conditions change. Costs increase. Changes beget changes. Humans err. Accidents happen. Materials fail. Financial problems arise. Applicable laws, regulations, and conditions of approval are misunderstood or ignored. For many reasons, property owners, developers, and operators may be engaged in land-use activities that vary in some material respect to the ones that were proposed and approved.

A tool for planners and planning officials to minimize inequitable results from informational constraints is to apply the precautionary principle when evaluating information about proposed projects.

Low-income and minority communities are especially vulnerable to bear the adverse impacts of these seven informational constraints—and those adverse impacts of land-use projects not fully understood and addressed. One reason for this vulnerability is the limited access that low-income people and people of color may have to good information and to experts who can detect weaknesses in the information being presented. Moreover, low-income people and people of color may be skeptical about, intimidated by, or overwhelmed by technical information and data about the environmental impacts of proposed land uses and the potential risks they pose. They may have very little familiarity with land-use approval processes and be reluctant to question people whom they perceive to be experts or people in power. They may doubt that the process can ever be fair or that any information from governments, businesses, or industries will ever be accurate or reliable, and therefore they may not differentiate between reliable and unreliable information. Likewise, there may be little productive common ground between pessimistic project opponents and optimistic project proponents. Finally, low-income people of color may lack the time, expertise, and resources to monitor compliance with applicable laws and conditions of approvals, and to seek enforcement against noncompliance. In the end, low-income and minority communities may bear harms that planners and land-use officials never intended when approving proposed land uses.

A tool for planners and planning officials to minimize inequitable results from informational constraints is to apply the *precautionary principle* when evaluating information about proposed projects. The precautionary principle states in essence that regulators do not need scientific certainty that a particular course of action will have adverse consequences in order to take action to prevent the action if it poses a substantial risk of adverse consequences. In other words, in the face of uncertainty, caution to prevent harm is to be preferred over attempts to remedy or address harm after it has occurred. The precautionary principle can be a mechanism for protecting low-income and minority neighborhoods from serious but inadequately understood risks.

Nonetheless, the precautionary principle should not be confused with an unrealistic policy goal of eliminating all risks from land-use projects. Where land-use decision makers face uncertain or unreliable information about the application of environmental justice to land-use issues and are faced with these issues numerous times, they may want to create an *environmental justice risk assessment and prevention initiative* to improve the information about the potential risks associated with certain land uses and to identify ways of preventing or minimizing those risks. For example, assume that a low-income neighborhood of color was experiencing a proliferation of automotive repair shops, truck storage yards, and car dealerships. The adverse environmental and land-use impacts of these facilities may be uncertain. Although the precautionary principle would urge caution in approving conditional use permits and variances for these uses in the absence of relatively thorough and reliable information, planning officials might also want to pursue an environmental justice risk assessment and prevention initiative for motor vehicle facilities. The purposes of this initiative would be to study the adverse impacts of facilities that store, maintain, repair, or operate motor vehicles in significant numbers and to identify ways of preventing, minimizing, or mitigating those impacts. Ideally, better information would result in better decisions that ensure environmental justice without unduly prohibiting land uses that do not adversely affect low-income and minority communities.

EXACTIONS AND IMPACT FEES (THINKING BEYOND STREETS AND SCHOOLS)

A not-so-obvious tool that could be part of a land-use planning model of environmental justice is the local government imposition of exactions (i.e., conditions) on approvals of industrial and commercial development near residential areas. Exactions require the developer to provide the public either real property (i.e., land, facilities, or both) or monetary fees as a condition for permission to use land in ways subject to government regulation (Been 1991; Rohan 1998, Section 9.01, 9-4 to 9-5). These dedications and fees provide the public facilities made necessary by new development, including schools, parks and open space, roads, sidewalks, public utilities, fire and police stations, low-income housing, mass transit, day care services, and job training programs.

There are five basic types of commonly imposed exactions:

1. On-site dedications, which consist of land and facilities within the developer's subdivision that the developer dedicates to the public

2. Off-site dedications, which consist of land and facilities outside the subdivision, yet dedicated by the developer

3. Fees-in-lieu-of-dedication, which are money contributions for the public provision of facilities that the developer otherwise would be required to dedicate

4. Impact fees, which capture from the private developer the public's costs of local capital-infrastructure and public-services needs caused by the development's impacts and

5. Linkages, which are facilities and/or fees provided by central-city commercial and industrial developers for the services necessitated by their specific development activities (Been 1991)

Cities and counties use exactions extensively, determining the amount demanded "either according to a nondiscretionary, predetermined schedule, or through case-by-case negotiations" (Been 1991, 481). They usually impose exactions during the subdivision map approval process because new subdivisions are significant sources of population growth that create the demand for additional public facilities and services. Other zoning approvals, such as rezoning or conditional use permits, may also trigger the expectation of exactions.

Exactions potentially benefit low-income and minority neighborhood residents in two ways. First, if a city or county requires a developer of a new residential subdivision to provide or pay for streets, parks, schools, public utilities infrastructure, and the like, the costs are borne ultimately by the residents

A not-so-obvious tool that could be part of a land-use planning model of environmental justice is the local government imposition of exactions (i.e., conditions) on approvals of industrial and commercial development near residential areas.

Local land-use officials seeking to impose exactions on industrial and commercial development and LULUs should do studies on the impacts of these developments or attempt to specify the development's direct and indirect impacts on the neighborhood.

(i.e., new homeowner) of the subdivision, not the general tax base. Therefore, residents of existing low-income or minority neighborhoods are not contributing taxes to infrastructure frequently enjoyed by upper-income whites in new suburban subdivisions. Furthermore, local tax revenues are not being diverted from services and facilities that support inner-city neighborhoods.

Second, government agencies can use exactions to mitigate the environmental impacts of new or expanding development in low-income and minority areas. Already, various federal, state, and local environmental regulatory programs require developers to dedicate land or pay fees to mitigate the environmental impacts of development in ecologically sensitive areas. A comprehensive environmental justice land-use program, though, might include environmental impact fees and dedications for inner-city industrial and commercial development. The exactions would be based on the various environmental and social impacts of intensive uses and LULUs on the surrounding neighborhood(s), not just the publicly funded local infrastructure, and would be earmarked for ameliorating amenities in the affected neighborhood(s). For example, an unsightly industrial facility might have to dedicate land for parks and open space, or to pay fees for these features. Similarly, an operator of a proposed waste facility might be required to contribute to a fund to be used for monitoring pollution levels and resident health status, as well as future cleanups of contamination related to the facility. An exactions program would be most attractive to an environmental justice approach to land-use regulation when either: 1) the local residents would not oppose the proposed land use if its adverse impacts were mitigated, or 2) complete prohibition of the proposed land use is politically or legally infeasible. The program, though, could apply only to new development or new activities (e.g., changes in existing uses) requiring development permits. In addition, it could not be used "to remedy existing infrastructure deficiencies, or to provide for operation and maintenance of facilities" (Rohan 1998, Section 9.01, 9-5).

Finally, the exactions program must be tailored to the impacts of the proposed developments. To survive a challenge under the Takings Clause of the Constitution, an exaction must bear an "essential nexus" to the legitimate government interest that forms the basis for regulating the development (*Nollan v. California Coastal Commission* 1987). It must also be roughly proportional in nature and extent to the impact of the proposed development (*Dolan v. City of Tigard* 1994). This two-part test applies to all land or facility dedication requirements and those impact fees imposed on an individualized, or ad hoc, basis (*Ehrlich v. Culver City* 1996).

There is lingering uncertainty about whether the *Nollan* "essential nexus" and *Dolan* "rough proportionality" requirements apply to legislatively adopted, formula-driven impact fees (*Amoco Oil Company v. Village of Schaumberg* 1995). The *Nollan* and *Dolan* standards appear to meet or exceed separate state constitutional tests requiring either a "reasonable relationship" or "rational nexus" between the exaction and the state interest in regulating the impacts of the development (Dana 1997). However, a few state courts require exactions to be tailored to impacts that are "specifically and uniquely attributable" to the proposed development, which is a higher standard than *Nollan* and *Dolan* (Dana 1997). In any event, local land-use officials seeking to impose exactions on industrial and commercial development and LULUs should do studies on the impacts of these developments or otherwise attempt to specify, preferably in quantitative terms, the development's direct and indirect impacts on the neighborhood. These studies would support arguments that the conditions are properly tailored to the government interest in regulating adverse impacts of development. In addition, as noted above, localities must avoid using exactions to remedy existing or past development impacts.

A CASE STUDY: THE USE OF OVERLAY ZONES AND REZONING, AUSTIN, TEXAS

The residents of East Austin are primarily African-American and Hispanic. Industrial, commercial, and residential uses are interspersed throughout the neighborhood. The City of Austin planned the area in 1928 as a "negro district" that would host most of Austin's industrial uses next to housing for African-Americans. Because local zoning allowed industrial uses on many parcels in East Austin, few or no obstacles existed to the siting of the noxious land uses that ended up there (e.g., at least two trash recycling plants, a power plant, a gasoline tank farm, and industrial facilities that use and emit hazardous and toxic substances).

Responding to complaints by neighborhood residents about specific land uses and the overall pattern of industrial zoning, the City of Austin conducted a study in 1997 showing that the area has a significantly higher percentage of industrial zoning than other areas of the city. Maps of land-use patterns in two East Austin neighborhoods—East Cesar Chavez and Holly—are shown in Figures 4-2 and 4-3.

The zoning report complemented an earlier study showing higher usage of hazardous substances in East Austin than in other areas of the city. Neighborhood residents, organized by PODER (People Organized for Defense of Earth and her Resources), demanded reform of the area's zoning, and the city council responded with three types of reform.

The first was the passage of an ordinance designating a large area of East Austin as the East Austin Overlay Combining District. The ordinance required a conditional use permit in the overlay district for any land use needing a hazardous materials permit from the Austin Fire Department, and any of the 14 land uses that the zoning code had allowed by right in the ordinance:

1. Agricultural sales and services (except nurseries)

2. Basic industry

3. Construction sales and services

4. General warehousing and distribution

5. Kennels

6. Light manufacturing

7. Limited warehousing and distribution

8. Recycling center

9. Resource extraction

10. Vehicle storage

11. Building maintenance services

12. Laundry services (except where the proposed use is 5,000 square feet or less)

13. Equipment sales

14. Equipment repair services

The ordinance did not change the underlying zoning designation of any parcel. However, new industrial or commercial uses or changes to existing industrial or commercial uses in East Austin, if falling within the list of conditional uses, would require a permit from the planning commission under zoning procedures designed to give local residents an opportunity to study and express their views on and object to the proposed uses. These procedures included notification of property owners and registered neighborhood associations living within 300 feet of a proposed site plan, and a public hearing at which concerned neighborhood residents could speak. The ordinance also contained a requirement that city staff report annually to the city council

SOURCES FOR EAST AUSTIN, TEXAS, CASE STUDY

- Greenberger 1997
- Haurwitz 1997
- Moscoso & Haurwitz 1997
- VanScoy 1997
- Austin, Texas Ordinance 970717-F
- City of Austin 1997
- City of Austin Planning, Environmental and Conservation Services Department 1997
- National Academy of Public Administration 2003, 89-116
- City of Austin Planning Department website.

Figure 4-2. *East César Chávez Neighborhood Planning Area*

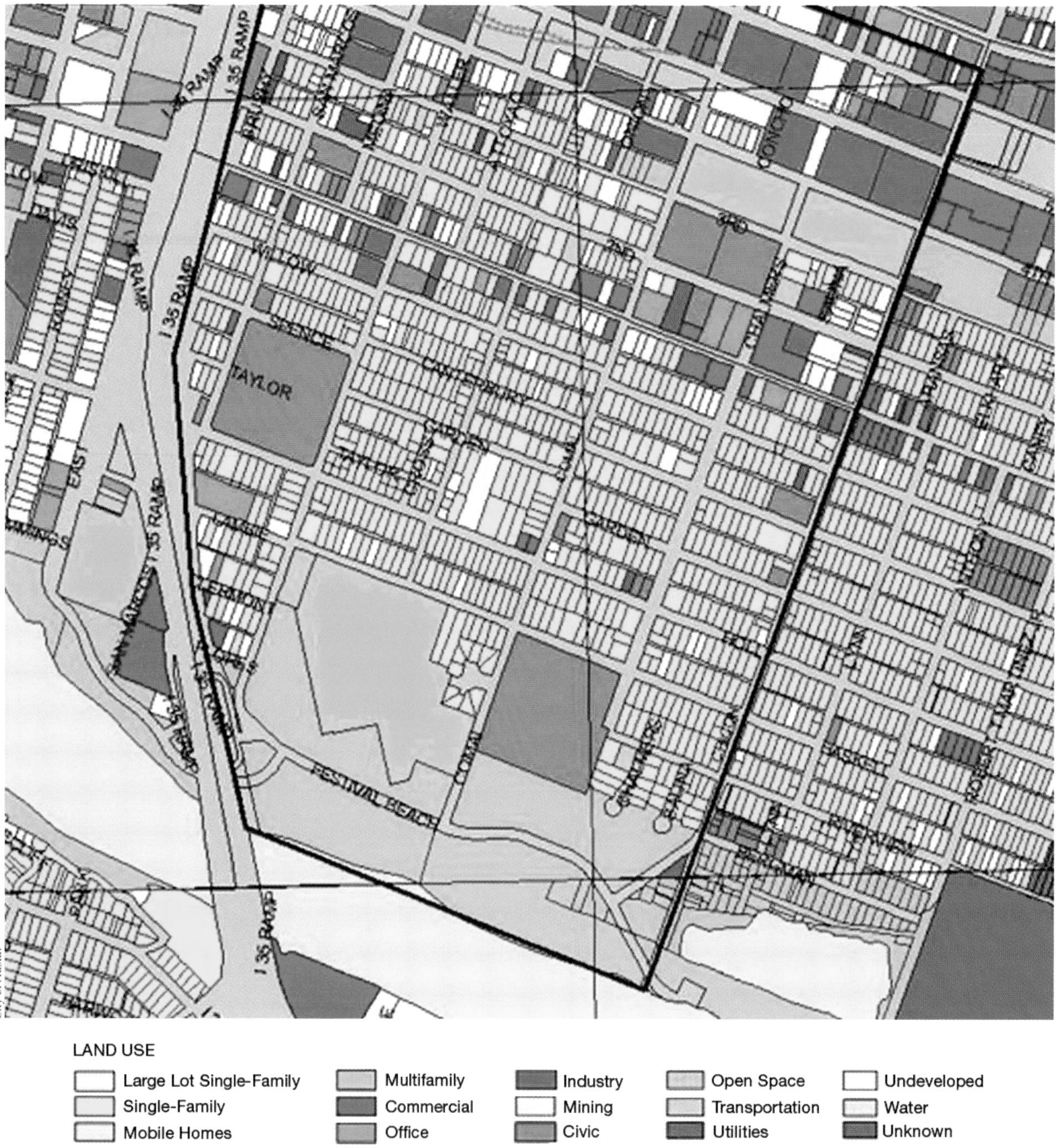
I 35 RAMP
TAYLOR
WILLOW
SPENCE
CANTERBURY
FESTIVAL BEACH
CONCHO
LAND USE
Large Lot Single-Family
Single-Family
Mobile Homes
Multifamily
Commercial
Office
Industry
Mining
Civic
Open Space
Transportation
Utilities
Undeveloped
Water
Unknown

City of Austin

about both the impact of the ordinance on the local neighborhood (i.e., the number of conditional use permits approved and denied, the change in the number of residential units constructed in the area, and other factors related to quality of life and the environment), and the impact of the ordinance on the property interests of industrial and commercial landowners (i.e., the change in the total appraised value of all affected development and other factors related to economic development and employment opportunities). The ordinance established the East Austin Overlay Combining District:

> **Section 25-2-169. East Austin (Ea) Overlay District Purpose and Boundaries.**
>
> The purpose of the East Austin (EA) overlay district is to reduce the concentration of intensive commercial and industrial uses in close proximity to residential

Figure 4-3. Holly Neighborhood Planning Area

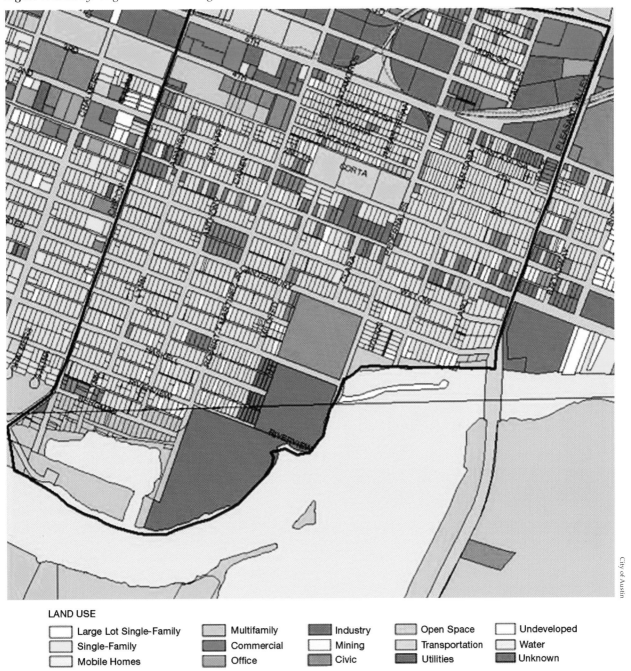

LAND USE

☐ Large Lot Single-Family	☐ Multifamily	☐ Industry	☐ Open Space	☐ Undeveloped
☐ Single-Family	☐ Commercial	☐ Mining	☐ Transportation	☐ Water
☐ Mobile Homes	☐ Office	☐ Civic	☐ Utilities	☐ Unknown

areas in East Austin and to mitigate the effect of commercial and industrial uses on nearby residential uses.

Except as provided in Subsection (C), the EA overlay district applies to property located in the area bounded by Interregional Highway 35, Airport Boulevard, and Town Lake.

The EA overlay district does not apply to land included in a neighborhood plan combining district. (Source: Section 13-2-190; Ordinance 990225-70; Ordinance 000406-81; Ordinance 031211-11)

Section 25-2-645. East Austin (EA) Overlay District Use Restrictions

This section applies to a use in the East Austin (EA) overlay district.

A use in a community commercial (GR), general commercial services (CS), commercial—liquor sales (CS-1), or limited industrial service (LI) base district is a conditional use if, under Section 25-2-491 (Permitted, Conditional, and Prohibited Uses) of the City Code, the use is permitted in the district and not permitted in a neighborhood commercial (LR) base district.

A medical office (not exceeding 5,000 square feet of gross floor area) use is a conditional use in a GR, CS, CS-1, and LI base district.

A service station use is a conditional use in a GR, CS, CS-1, and LI base district.

A guidance services use and a communication service facilities use are conditional uses in all base districts.

A pawn shop services use is prohibited in a GR, CS, CS-1, and LI base district. (Source: Section 13-2-191; Ord. 990225-70; Ord. 990520-70; Ord. 990520-70; Ord. 031211-11)

PODER (People Organized for Defense of Earth and her Resources) successfully campaigned to get Austin to rezone a neighborhood affected by environmental injustice. Here, PODER's Young Scholars for Justice (YSJ) speak out standing outside the site of their most current battle: the city's Holly power plant.

Second, the city council rezoned individual parcels from industrial to either commercial or residential uses. For example, the city council rezoned the site of the BFI recycling plant, which has posed problems of blowing trash, rats, noise, traffic, and a five-alarm fire, from limited industrial to limited office, and the site of the Balcones recycling plant, which had caused neighbors to complain about aesthetics, noise, and traffic, from limited industrial to residential. The rezoning did not automatically shut down the existing uses of these properties, but it prevented expansion of their uses or any new industrial uses unless the new owner were to resume the exact same land-use activity within 90 days. Furthermore, the city council rezoned a number of lots containing residences to residential designation so that they could not be converted to industrial or commercial use.

Third, the city engaged in neighborhood-based planning in East Austin, listening to residents' concerns and ideas for their neighborhood. In addition, a member of PODER from East Austin was elected to the city council and another member of PODER from East Austin was appointed to the planning commission. A "Neighborhood Academy" was created to build the capacity and knowledge base of neighborhood residents to engage in planning for their communities. The city developed a set of plans for East Austin that call for reducing industrial uses and for designating most parcels for residential, public, and mixed uses that preserve the residential integrity of the neighborhoods in East Austin. Maps of the future land-use plans for two East Austin neighborhoods—East Cesar Chavez and Holly—appear in Figures 4-4 and Figure 4-5.

Much of the 1997 East Austin Overlay Combining District ordinance was replaced with a more comprehensive set of zoning that implements the plans, but an East Austin Overlay Combining District remains in place. The neighborhood plan for East Austin can be viewed at www.ci.austin.tx.us/zoning/central_east_austin.htm, and the city ordinance is available at http://www.amlegal.com/austin_nxt/gateway.dll/Texas/austin/title25landdevelopment/chapter25-2zoning?f=templates$fn=altmain-nf.htm$3.0#JD_25-2-169.

Figure 4-4. *The East César Chávez Future Land Use Plan*

Neighborhood Gateway · Multi-Family · Residential · Civic
Mixed Use · Openspace · Historic District

City of Austin

Figure 4-5. Holly Neighborhood Planning Area Future Land Use Map

LAND USE CATEGORIES

Single-Family   Commercial   Office   Civic

Multifamily   Mixed Use   Industrial   Open Space

CHAPTER 5

Community Participation

H aving meaningful opportunities for all persons to participate in the environmental and land-use decisions that affect
them is a core principle, not only of environmental justice but of
land-use planning, as discussed in Chapter 2 of this PAS Report.
Paul Davidoff (1965, 332) wrote:

> If the planning process is to encourage democratic urban governance
> then it must operate so as to include rather than exclude citizens from
> participating in the process. "Inclusion" means not only permitting
> the citizen to be heard. It also means that he be able to become well
> informed about the underlying reasons for planning proposals, and
> be able to respond to them in the technical language of professional
> planners.

Sherry Arnstein (1969) called for public involvement that involves true citizen power over land-use decisions, not merely token placation, consultation, or information, or other nonparticipatory interactions with government officials. Her ladder of citizen participation appears in Table 5-1.

Planners and planning officials have a multitude of possible ways to facilitate meaningful participation by low-income and community residents in land-use planning and decision making. A sound environmental justice policy makes use of as many ways to encourage and to support community participation as possible, and does not rely on just a few participatory tools. For further details on designing and implementing participatory processes, see Randolph 2004, 47, 55–74; American Planning Association 2006, 46–67; California Governor's Office of Planning and Research 2003, 142–48; Kelly and Becker 2000, 111–29; Camacho 2005; Martz 1995.

TABLE 5-1. A LADDER OF CITIZEN PARTICIPATION

Level	Type of Participation	Degree of Participation
8	Citizen control	Citizen power
7	Delegated power	Citizen power
6	Partnership	Citizen power
5	Placation	Tokenism
4	Consultation	Tokenism
3	Informing	Tokenism
2	Therapy	Nonparticipation
1	Manipulation	Nonparticipation

Source: Arnstein 1969, 217.

WHAT IS NECESSARY

Commitment

Involving low-income and minority community residents in land-use planning and decision making first requires a clear, supportable commitment to doing so:

- Make formal commitments to widespread, effective participation of all community members in planning and zoning.

- Plan for staff time and resources to facilitate public participation.

- Develop benchmarks and performance assessments for increased participation and opportunities for participation.

- Inform, and invite the support of, elected officials, appointed officials, staff managers, and other departments about enhancing the effective participation of residents in low-income communities of color.

- Seek the appointment of members of low-income and minority communities to planning and zoning boards, as well as boards of appeal or adjustment and regional planning entities; effective participation and real commitment to participation mean that residents of low-income and minority communities are involved in the exercise of local land-use planning and regulatory power as members of decision-making bodies.

Relationship Building

Planning staff must develop positive and ongoing working relationships with residents of low-income and minority communities if they expect participation to be effective and meaningful:

- Identify and contact neighborhood organizations and community leaders in low-income and minority communities.

- Develop working relationships and partnerships with grassroots environmental justice groups and neighborhood groups in low-income and minority communities.

- Avoid paternalism and patronizing attitudes; strive to be a partner with local citizens in planning the locality's future.

Information

The flow of helpful and reliable information on a regular basis from planners to residents and from residents to planners builds relationships, mutual commitment, and further communication. The information should be communicated in ways understandable to community residents, including generous use of visual presentation, graphic presentation of data, and translation into the dominant language of members of the community. Some specific actions that can be taken include the following:

- Create a general community suggestion or community input system (e.g., mail box, e-mail address, Internet site, etc.) that allows public input on problems, issues, and ideas not currently on a decision-making body's agenda.

- Use local residents in gathering data and reporting violations of land-use laws (e.g., "bucket brigades" in which community residents test and monitor air quality conditions with a relatively inexpensive but industry-standard air-sampling device in a five-gallon bucket (Pastor and Rosner 2002)).

- Host communitywide planning input drop-by sessions at a central location (e.g., a sports arena, shopping mall, high school, major park, library, local fair, etc.), where residents can conveniently stop to register input on proposed plans or alternatives presented on displays through visual preference surveys (and perhaps also pick up a free hot dog or donut—food has a way of facilitating participation).

- Use community meetings in neighborhoods to both share information and listen to concerns; establish dialogue and participatory deliberation.

- Publish sources of information, including notices of meetings and information on how to participate, in the dominant languages of significant numbers of local residents.

- Dedicate portions of information sources (e.g., newsletters, websites) to environmental justice issues.

- Publish an easy-to-read guidebook or pamphlet on public participation.

- Disseminate information broadly, including through public workshops, newsletters, postal mailings, notices distributed in utility bills or at public schools, a speaker's bureau, radio and television broadcasts, electronic mailings, Internet websites, and similar computer-based information networks.

- Conduct surveys of community residents, as well as business owners, operators, and employees in the area.

- Use visual images whenever possible.

- Present visual surveys of communities facing environmental justice issues through PowerPoint or other electronic display, an Internet page, or photographs.

- Host physical tours of communities facing environmental justice issues for community residents and other interested parties (including business, industry, and development representatives).

The 2006 APA Davidoff award recipient for advocacy planning is Marva Smith Battle-Bey, who has engaged low-income and minority community residents of South Los Angeles in community-based economic development and neighborhood planning.

- Use a geographic information system (GIS) to present information about existing conditions and proposed plans, and to engage community residents in neighborhood visioning and planning, including simulation and modeling of various scenarios created with community input.

- Host open "conversations" between community residents and existing industry and business representatives to share in an unstructured way their respective interests and concerns; establish ground rules to focus the dialogue on sharing concerns rather than making accusations or assessing blame.

- Create a "planning academy" to provide information and training to community residents on planning and land-use regulatory issues.

- Use translators for meetings affecting people for whom English is not their dominant language and translate agendas, minutes, and major documents into their dominant language.

- Conduct physical tours of communities facing environmental justice issues so that planning staff and local official can better understand the issues.

- Do environmental justice audits, environmental impact assessments, and/or community impact assessments that:

  1. provide information to community residents and other affected persons;

  2. involve community residents and neighborhood groups in gathering data;

  3. are printed in the dominant languages of community residents;

  4. use language comprehensible to the layperson without oversimplifying; and

  5. serve as useful tools in planning and decision making.

Open and Accessible Government

An environmental justice participatory strategy ensures that government decision-making venues are fully open and accessible to residents of low-income and minority communities, and that public officials are responsive to all members of the public:

- Hold planning commission meetings and similar land-use meetings at times and on days that members of low-income and minority communities can attend (e.g., evenings, weekends), and hold meetings on major planning, zoning, or projects approval decisions in the affected neighborhood/area.

- Use translators for meetings affecting people for whom English is not their dominant language, and translate agendas, minutes, and major documents into their dominant language.

- Publish sources of information, including notices of meetings and information on how to participate, in the dominant languages of significant numbers of local residents.

- Provide accessible opportunities for community members to comment on drafts of plans, proposed zoning, project/permit applications, and environmental impact assessments through written comments, e-mail comments, phone messages, and oral comments at advertised public meetings.

- Respond to comments in hearings, reports, final documents, and/or decisional records.

- Ensure that all meetings of decision-making bodies and appointed advisory boards are advertised and open to the public, except when expressly exempted by state open meetings laws (e.g., executive sessions to discuss pending litigation, personnel matters, etc.).

- Limit decision makers' discussion of projects and proposals to open public meetings; be proactive in discouraging deliberations and decision making in off-record conversations.

- Involve community residents early in decision making about planning, zoning, permit decisions, public infrastructure, and the like; do not wait until plans are well developed or essentially completed.

- In general, respond to public input whenever and however it is received.

NEIGHBORHOOD-BASED PLANNING

A participatory approach involves community residents in planning activities that identify their goals, aspirations, and visions for their neighborhood and its surrounding environment. Community residents are sources of information about land-use and environmental problems, as well as community assets. A specific community's self-identified needs and goals should also be as much a part of any land-use plan for that community as the needs and goals of the broader community (i.e., the city, county, or region). Moreover, consensus among residents about a plan for their neighborhood produces the buy-in necessary for planning to be effective in the long run. A variety of techniques for engaging in planning with neighborhood residents are readily available:

- Host physical tours of communities facing environmental justice issues for community residents and other interested parties (including business, industry, and development representatives).

- Present visual surveys of communities facing environmental justice issues through PowerPoint or other electronic display, an Internet page, or photographs.

- Conduct surveys of community residents, as well as business owners, operators, and employees in the area.

- Seek and facilitate community or neighborhood visioning exercises with a diverse range of community residents; visioning should identify important community values and goals.

- Use planning and design charrettes in meetings with community residents (a charrette is a multiday, intensive, interactive, interdisciplinary design or problem-solving process that identify issues and needs, generate ideas, and develop strategies and specific implementation methods).

Charrettes are valuable means of providing meaningful, deliberative participation of community residents in development of specific plans and designs. The National Charrette Institute process (shown in the graphic below) involves cycles of informed input and feedback.

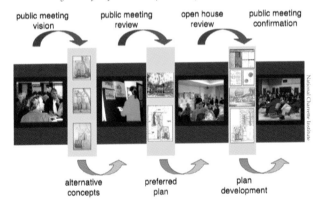

MAPPING AND VISIONING IN THE LITTLE VILLAGE COMMUNITY, CHICAGO, ILLINOIS

Residents of the Little Village neighborhood are combining technology-aided spatial analysis and community participation to address the problems and opportunities of their low-income, mostly Latino neighborhood in the South Lawndale neighborhood in Chicago. The Little Village Environmental Justice Organization (LVEJO) received funding first from the University of Illinois-Chicago and then from the Northeastern Illinois Planning Commission to use GIS technology to engage in community-based mapping, visioning, and planning.

Initially, college and high school youth in the neighborhood worked with a geography graduate student to: 1) learn the basics of GIS; 2) perform a block-by-block inventory of 130 blocks in the area; 3) enter the data; and 4) map the area by property characteristics (including land use), health data, and census tract data. The data and the initial engagement of community residents in data collection subsequently led to the following developments:

- An information technology (IT) training program for both adults and youth
- Community mapping and creation of an inventory of assets
- Training of leaders in the use of hand-held computers and digital cameras for inventorying conditions
- Skill development in group facilitation
- Learning how to do consensus building, using English and/or Spanish
- Training in methods used in planning and development, and in methods of participatory planning, including charettes

Little Village Environmental Justice Organization

- Public forums in schools, churches, and community group to begin the participatory planning process
- Cooperation with housing developers in identifying possible housing development projects
- Co-participation with schools and city agencies to plan and develop a high school, three school gardens, and a campus park, including several months of dozens of meetings, guided tours, and the use of focus groups and a charette design process
- Multiparticipant construction of a park and playground at the Little Village Boys and Girls Club
- A plan for a plastics recycling facility in an industrial park, resulting from "adults and youth doing a walking tour of the property, taking pictures, studying City of Chicago Department of Planning Base Maps, creating drawings, writing up a formal proposal to the City of Chicago Department of Planning, [and] presenting the plan to the Planning Commissioner for Industrial Development and at eight public meetings where over 2,000 community residents were in attendance"
- Establishment of two new bus routes and participation on a transit authority community committee
- Planning meetings with community residents and city planning officials about open space, brownfields, parks, and recreation facilities
- Block-by-block photo documentation followed by discussions among community residents about problems, assets, opportunities, goals, and possible solutions
- Community forums to view and discuss GIS maps and other data, and to begin the visioning process

LVEJO believes that both representative participation and technology are critical to creating a sustainable community:

> The foundation of a sustainable community is the active participation of those who live and work there in the planning, implementation and evaluation of projects. . . . One phase in sustainable development is for community residents to be able to "see" their entire community and all its parts. Spatial analysis is a powerful tool that, when combined with ongoing participatory methods, allows for neighborhood members to plan for their future. The various pieces of communities . . . need to be viewed by communities not just in bits and pieces as is the usual case but in a holistic framework. ▪

Source: Little Village Environmental Justice Organization website, www.lvejo.org, accessed December 31, 2006

1,200 trucks per day travel from the Port of San Diego's Marine Terminal through the community of Barrio Logan, spewing carcinogenic diesel particulates. The Environmental Health Coalition of National City, California, is working with community residents and the Port to develop a Clean Port Plan to reduce toxic air emissions.

- Use interactive games, map-drawing exercises, picture-drawing exercises, and other similar activities with area residents.
- Use focus groups to identify and discuss particular issues.
- Use advisory boards and task forces from the community.

See Chapter 3 of this PAS Report for additional descriptions of neighborhood-based planning. For an example of participatory neighborhood-based planning, led by the Little Village Environmental Justice Organization on Chicago's West Side, see the sidebar.

CONFLICT AVOIDANCE OR RESOLUTION

Planners can play a role in facilitating cooperation between low-income and minority community residents and the businesses and industries in their areas or developers or proponents of new projects in their areas. Participation in land-use decision making encompasses participation not only in formal hearings and official meetings but also in mediations, negotiations, and collaborative problem-solving efforts. Some specific actions that can be taken include the following:

- Require developers and project proponents to meet with residents in the affected community prior to filing an application for the development.
- Create multi-stakeholder, collaborative, problem-solving groups or task forces, using negotiation and/or mediation techniques to address particular problems or conflicts.
- Host open "conversations" between community residents and existing industry and business representatives to share in an unstructured way their respective interests and concerns; establish ground rules to focus the dialogue on sharing concerns rather than making accusations or assessing blame.

CHAPTER 6

The Environmental Impact Assessment as a Tool for Implementing Environmental Justice

An assessment of the environmental impact of land-use proposals and patterns is an important tool in achieving clean, healthy, vibrant, and just communities for all people, including low-income people and people of color. Such assessments allow decision makers to evaluate whether new policies and plans are needed, to determine whether specific projects and land uses will have adverse impacts on low-income people and people of color, and to select appropriate minimization and mitigation conditions for projects that may be appropriate if redesigned or conditioned. Planners face three issues in employing environmental impact assessment as an environmental justice tool: 1) the types of decisions that warrant environmental impact assessment; 2) the types of impacts to assess; and 3) the use of the information gathered from the environmental impact assessment.

WHEN IS ENVIRONMENTAL IMPACT ASSESSMENT APPROPRIATE?

The broadest use of environmental impact assessment is to evaluate the environmental impacts of current land-use patterns and practices to determine whether new land-use plans, policies, or regulatory reforms are needed. For example, a locality may wish to evaluate the environmental impacts of industrial uses in and near residential neighborhoods to determine how these land-use patterns are affecting environmental conditions, human health and safety, neighborhood vitality, and the economy.

The next broadest use of environmental impact assessment is to evaluate government policies, plans, and programs for their environmental impacts. For example, a locality may be planning to focus future residential development and growth in a specific part of the jurisdiction. It may be considering a comprehensive industrial revitalization and reinvestment policy. It may be formulating a specific road-widening project or a jurisdictionwide plan to underground utilities. It may be addressing stormwater runoff patterns in the jurisdiction. It may be seeking to build an affordable housing project and a new school on a brownfield site. In each of these examples, the potential for significant impacts on the environment exists. The potential that the projects, plans, or policies will have significant impacts on low-income and minority communities also exists.

Next, local governments may conduct an environmental impact assessment when considering discretionary land-use approvals (e.g., rezoning, conditional use permit, variance, subdivision or PUD approval) for any proposed project or land use that reasonably might have a significant impact on the environment. Some states, like California, require local and state government agencies to conduct this assessment, a requirement that may be vigorously enforced by courts. In other states, the statutory requirements are more lax, or the choice to conduct such an assessment as part of the land-use approval process may be discretionary. Making environmental impact assessments a standard practice for projects open to discretionary approval can be a good idea because such practice will provide information to decision makers about the propriety of the use, particularly its compatibility with the surrounding area and with the public health, safety, morals, and welfare. It also ensures that environmental impacts are being evaluated for all projects, regardless of the racial or socioeconomic composition of the communities in which they may be located. Having a standard practice does not dictate the breadth and depth of the assessment, though; it can range from merely completing a checklist of possible impacts with further study of those impacts likely to be significant to an in-depth, thorough analysis of all potential impacts. In addition, the trigger for engaging in the assessment may be broader, such as projects that reasonably *might* have *any* environmental impacts at all, or narrower, such as projects that are reasonably *likely* to have *major* environmental impacts.

Local governments, however, might conduct environmental impact assessments only for certain types of land uses or projects, such as all waste facilities or all industrial facilities. The local government may wish to develop a zoning code provision listing uses likely to have significant environmental impacts and requiring environmental impact assessments in those circumstances.

It may be the practice of local governments to conduct environmental impact assessments on an ad hoc basis, when a planner or administrator is concerned about the impacts of a particular project, when government decision makers request analysis of a project's impacts, or when members of the public who are opposed to the project request additional information.

Finally, local officials may choose to perform environmental impact assessments only for projects in particular areas currently exposed to or at risk of disproportionate environmental impacts. The specific areas selected for the

The broadest use of environmental impact assessment is to evaluate the environmental impacts of current land-use patterns and practices to determine whether new land-use plans, policies, or regulatory reforms are needed.

performance of environmental impact assessments might be selected on the basis of: 1) the racial, ethnic, and/or economic make-up, using demographic data, or 2) existing environmental and land-use conditions that: a) exist in a higher quantity or concentration than occurs in other areas of the local jurisdiction, or b) cause harms and risks of harm to human health and safety, the vitality of the natural environment, property values, or neighborhood integrity and character.

With respect to the first option, areas with a high percentage of racial or ethnic minority population and/or a high percentage of people below the poverty line would seem to be appropriate areas for environmental impact assessment. To avoid litigation based on the perception that certain areas are being given more favorable treatment on the basis of race, officials will need to document evidence that low-income and minority communities have historically suffered a higher proportion of environmental harms and intensive land uses, considering both local data and national studies.

Planners and local officials must consider environmental impacts in the context of the diverse characteristics of low-income and minority communities. Featured here (clockwise) are images of a community garden in North Philadelphia, mixed uses in a San Antonio barrio, a crowded San Francisco Chinatown street, and a home in Pittsburgh's Oak Hill, a mixed-income Hope VI development.

With respect to second option, numerous, rather simple methods are available to designate areas with poor environmental conditions and, therefore, as appropriate for environmental impact assessment. For example, planners might map:

- the location of Toxic Release Inventory (TRI) sites;

- the locations of hazardous waste transfer, storage, and disposal facilities (TSDFs);

- the locations of National Priority List Superfund sites or sites identified as contaminated by hazardous waste by state or local agencies;

- patterns of industrial zoning in close proximity to residential areas; and

- patterns of air pollutants across the local jurisdiction.

These maps would reveal many of the areas where environmental impact assessments would be appropriate.

TYPES OF IMPACTS TO ASSESS

The most obvious environmental impacts to measure are those on the physical environment. These include impacts on:

- air quality;

- water quality;

- water hydrology, flooding, and groundwater recharge;

- water supply;

- soils;

- erosion and siltation;

- pollution levels;

- exposure to toxic or hazardous substances, including waste;

- generation of waste and/or litter;

- biodiversity, wildlife, fish, and their habitat;

- climate;

- the consumption and waste of natural resources, including energy;

- geological conditions and exposure of people or structures to geological hazards;

- traffic levels and flow and impacts on the traffic load and capacity of streets and other roadways;

- pedestrian flow and safety;

- ambient noise levels;

- the geographic distribution of people, including growth inducement, growth location, and displacement;

- the quality and quantity of the housing supply;

- deterioration, waste, or abuse of physical structures and infrastructure;

- the quantity and quality of open space and recreational areas;

- aesthetic or visual conditions;

- sites of archeological, historical, or cultural value;

- emergency response and evacuation methods; and

- public health generally.

Planners have an increasing amount of information about the relationship between land uses and environmental conditions. For example, the California Environmental Protection Agency and the California Air Resources Board

(2005) have issued an especially helpful report on the relationship between air quality and land use. It contains assessment tools for evaluating the air quality impacts of existing and proposed land uses (including cumulative impacts), as well specific planning and land use regulatory options for addressing these impacts. (See Figures 6-1 and 6-2 for maps of San Diego showing the distribution of air toxics and the risk of cancer, with the greatest concentrations and risk in the low-income, minority neighborhood of Barrio Logan.)

Figure 6-1. *(Top) Estimates of Cancer Risk from Toxic Air Contaminants, Logan Area Communities*

Figure 6-2. *(Below) Toxics and Freeways, Logan Communities*

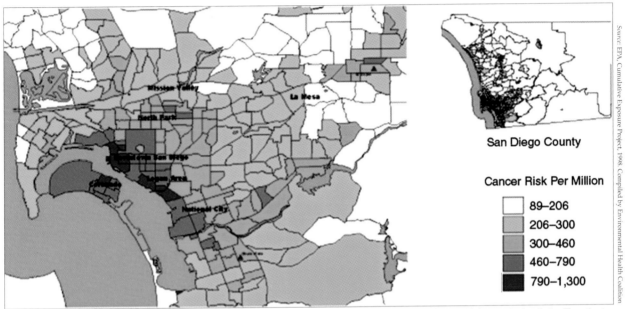

Source: EPA, Cumulative Exposure Project, 1998. Compiled by Environmental Health Coalition

Note: EPA estimates the concentrations of toxics in the air from all major sources, including industry and traffic, then estimated cancer risk due to toxic air pollution. The estimates are not considered highly reliable at the census tract level. For this reason, the data are presented here in ranges.

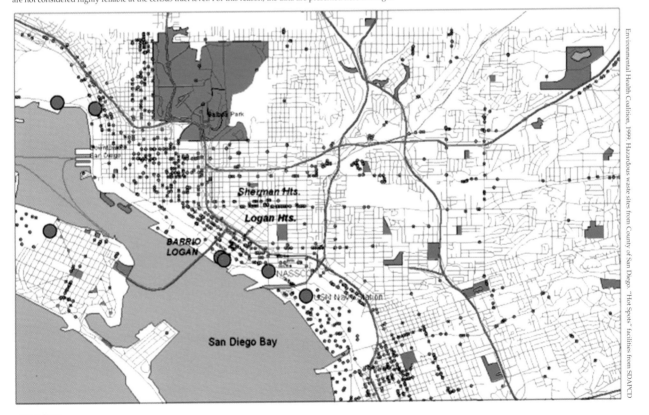

Environmental Health Coalition, 1999. Hazardous waste sites from County of San Diego; "Hot Spots" facilities from SDAPCD

Map Key

⋀ Freeways & streets within 900 ft. of freeway   ◎ Air Toxic Hot Spot Facilities   ● Hazardous Materials and/or Wastes

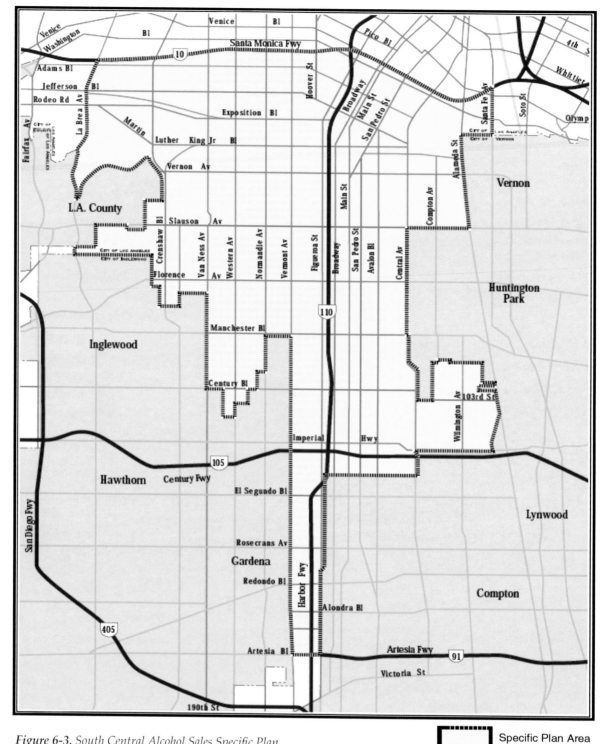

Figure 6-3. *South Central Alcohol Sales Specific Plan*

▭▭▭▭ Specific Plan Area

Planners and local officials may also wish to consider social and economic impacts of proposed land-use projects when completing environmental impact assessments (Foster 1999b). In fact, the line between the physical impacts and social impacts of land uses is blurry at best, and often the two categories of impacts are interrelated. Some of the most significant social impacts to consider include impacts on community character and identity, sense of place, the vitality of civic life, crime levels, economic opportunity, and fair treatment. For example, the City of Los Angeles identified the overconcentration of establishments selling alcohol in South Central Los Angeles as having an adverse impact on health and safety, crime, and neighborhood stability and character. In response, it adopted the "South Central Alcohol Sales Specific Plan" to require conditional use permits for the sale of alcohol beverages in the South Central Specific Plan Area (see Figure 6-3) only upon finding that the approval of an permit would not result in an undue concentration of such establishments in the area. (City of Los Angeles 1997)

A different type of environmental impact assessment that arises in the environmental justice context is the assessment of the distributional impacts of land uses and environmental conditions by race and class. Here, the focus is on whether the environmental impacts of a particular land use or project will fall disproportionately on low-income and minority communities in comparison to other communities. There is considerable disagreement about how to measure disproportionate impact, as recent experience in New York indicates. The Disproportionate Adverse Environmental Impact Analysis Work Group of the New York State Department of Environmental Conservation (New York State Department of Environmental Conservation 2004) failed to reach consensus about which of the six following methods should be used to assess disproportionate impact:

A different type of environmental impact assessment that arises in the environmental justice context is the assessment of the distributional impacts of land uses and environmental conditions by race and class.

1. Comparative community of concern analysis, which compares the demographics of affected areas with the demographics of a nonaffected area selected as a reference

2. Proportional impact analysis by demographics, which compares, within an area, the demographics of the subareas closest to the site with other subareas

3. Proportional impact analysis by project impact, which identifies zones of varying impacts within the affected area and analyzes the relationship between demographics and zones of impact

4. Alternative site analysis, which compares the demographics of the selected site with the demographics of alternative sties

5. GIS burden analysis, which compares the relationships between demographics and total environmental burdens in the project area with the relationships between demographics and total environmental burdens in a separate reference area

6. Burdened area analysis, which determines if the project area is one in which high levels of existing environmental burdens overlap with minority and low-income demographics. (New York State Department of Environmental Conservation 2004, 12–18)

Whether assessing environmental impacts, distributional impacts, or both, planners implementing environmental justice principles will give particular attention to cumulative impacts and synergistic impacts. "Cumulative impact, as defined by NEPA [the National Environmental Policy Act, is the sum of the incremental impact of a proposed action when added to other past, present, and reasonably foreseeable future actions" (Rechtschaffen 2003, p. 125). NEPA requires federal agencies to consider cumulative impacts when preparing environmental impact statements for proposed projects.

Consideration of cumulative impacts is particularly important in addressing environmental justice concerns because low-income and minority communities: a) have high numbers and concentrations of existing intensive land uses that individually contribute to degraded environmental conditions in their areas, and the potential for more such facility owners or operators to identify these areas for future development or expansion sites;

and b) "are exposed to multiple environmental hazards, through various pathways" (Foster 1999b, 270). A proposed industrial facility in an area with a high concentration of low-income and minority residents, for example, might not be anticipated to produce enough pollution by itself to cause a major risk to human health, but when combined with all the other existing industrial facilities in the area, the sum total of the pollution poses a major risk to human health.

The true measure of cumulative impacts is often more than the mere sum of the individual impacts (Collin and Collin 2005). There may be multiplier effects when quantities of, concentrations of, or exposures to environmental conditions increase. Different kinds of conditions may have synergistic effects when they interact with one another, creating impacts distinct in character, and considerably more harmful, than each separate condition. In addition, substances may bio-accumulate in organisms, having long-lasting and far-reaching impacts. An environmental justice approach to environmental impact assessment gives priority to studying the cumulative, multiplier, synergistic, and bio-accumulative effects of environmental conditions.

Different kinds of conditions may have synergistic effects when they interact with one another, creating impacts distinct in character, and considerably more harmful, than each separate condition.

SYSTEMATIC PROCESS

Randolph (2004) urges a systematic process for engaging in environmental impact assessment, which he terms EIA. His description of such a system follows:

> In conducting an EIA, it is important to assess the environment systematically. Generally, the assessment focuses on indicators of change. The following list defines impact variables or important components of the environment, indicators of change, and thresholds or standards for those indicators. . . .
>
> **Environmental Impact Variables, Indicators, and Thresholds**
> Impact Variables: Components of the environment that are important (e.g., water quality)
>
> Impact Indicators: Measures that indicate change in an impact variable (e.g., dissolved oxygen)
>
> Impact Thresholds or Standards: Values of impact indicators above or below which there is a problem; used to evaluate the impact (e.g., 5 ppm minimum of dissolved oxygen)
>
> EIA aims to predict future change in impact indicators that are likely to result from the proposed action. "With-Without" (W-W/O) analysis is used to do this. . . . The future change of a selected indicator is predicted with the proposed action and plotted on the graph. It is important to know the change that actually results from the action, so it is necessary to also plot the change in the indicator that would result if the action were not undertaken. The "without" line plots this change. The "impact" of the proposed action is the difference between the "with" and "without" lines, not the difference between the "with" line and today's value or baseline. . . .
>
> **The EIA Process**
> The following list gives an outline of a generalized EIA process. . . .
>
> *Scoping*: Design the process; draft the work program; identify issues, impact variables, parties to be involved and methods to be used.
>
> *Baseline Data Studies*: Collect initial information on baseline conditions and important impact variables, which may include socioeconomic as well as environmental parameters.
>
> *Identification of Impacts*: Concurrent with baseline studies, identify and screen impacts of alternative actions: variables, indicators, and thresholds.
>
> *Prediction of Impacts*: Estimate the magnitude of change in important impact variables and indicators that would result from each alternative using W-W/O analysis. Employ project outputs, simple algorithms, simulation models as needed.

Evaluation of Impacts and Impact Mitigation: Compare indicator impacts to thresholds; determine relative importance of impacts to help guide decisions; evaluate plans for mitigation of impacts.

Presentation of Impacts: Present impacts of alternatives in concise and understandable format." (Randolph 2004, 613–15)

CONCLUSION

The above sections briefly document how environmental impact assessments can be used in the context of environmental justice. Principles of environmental justice call for the involvement of affected local residents, especially in particularly vulnerable communities like low-income communities of color, in all stages of the environmental impact assessment process:

1. The initial, nondetailed assessment to determine if the potential impact reaches the threshold to prepare a study.

2. The definition of the scope of the study.

3. The data-gathering stage and the preparation of a draft study.

4. The evaluation of the adequacy of the draft and the commenting period.

5. The final study and its impact on land-use decision making (Foster 1999a, 195–200).

CHAPTER 7

Community Infrastructure, Housing, Redevelopment, and Brownfields

An environmental justice strategy addresses public-sector development, not just public-sector regulation of private-sector development. A comprehensive land-use policy of environmental justice integrates these two spheres. All too often, local governments attempting to address social equity choose to focus on either public infrastructure in low-income and minority neighborhoods or regulation of land uses in these neighborhoods, but not both. Public or public-private development projects coordinated with environmental justice planning and regulatory goals provide critical opportunities for local governments to improve proactively the environmental conditions in low-income and minority communities.

If communities are to thrive, though, they need attractive, safe, and functional places for their residents to gather, play, and build relationships with one another and with their social and natural environments.

The following six policies should shape public infrastructure development: 1) distributional equity and accessibility; 2) community capacity and vitality; 3) prevention of health risks and promotion of good health; 4) public participation; 5) conversion of underperforming assets into performing assets; and 6) policy integration and coordination. Moreover, specific environmental justice issues arise with the following four types of public development projects: 1) community infrastructure; 2) housing; 3) redevelopment; and 4) brownfields.

DISTRIBUTIONAL EQUITY AND ACCESSIBILITY

A fundamental principle of environmental justice is that public facilities and infrastructure should be distributed in low-income and minority communities in roughly the same numbers, quality, scope, and degree of accessibility as they are in other communities within the locality or region (California Governor's Office of Planning and Research 2003, 25-26; Rubin 2006). As discussed in Chapters 2 and 3 of this PAS Report, significant disparities in public infrastructure exist between many low-income and minority neighborhoods and other parts of their localities (Haar and Fesler 1986; Bond 1976; Garcia and Flores 2005; Harwood 2003). Even where facilities exist in low-income and minority communities, they are likely to be smaller, older, less well-maintained, and of poorer quality than others in the locality. Likewise, the locations of facilities combine with the lack of transportation options appropriate to the neighborhood's physical and socioeconomic context to prevent access to communitywide or regional facilities. The facilities lacking in adequate numbers in low-income and minority communities can range from parks, open space, and recreational areas, to retail establishments and restaurants with healthy menus, to sidewalks and public transportation, to refurbished or well-maintained streets, sewers, and stormwater systems, to schools and community centers with adequate space and equipment.

COMMUNITY CAPACITY AND VITALITY

Another fundamental principle of environmental justice is that low-income communities and communities of color require facilities, infrastructure, and public development policies that build and strengthen community capacity and vitality. Low-income and minority areas of a locality should not be seen as "problems" to be ignored, manipulated, or solved by expert planners. Instead, they should be seen as organic and dynamic communities with strengths and weaknesses, resources and needs, and opportunities and challenges. Public facilities and private facilities aided by public involvement should seek to maximize a community's strengths and resources while addressing its weaknesses and needs. Infrastructure planning should build on the existing assets of neighborhoods.

At a minimum, all communities need basic infrastructure to support the activities that occur within communities. If communities are to thrive, though, they need attractive, safe, and functional places for their residents to gather, play, and build relationships with one another and with their social and natural environments. These places need to be places for which the community residents have some sense of investment, ownership, or control. Public infrastructure plays a critical role in supporting and facilitating community vibrancy, and vibrant signs of life within the most disadvantaged neighborhoods within a locality or region contribute to the vitality of the overall locality and region. Moreover, low-income and minority neighborhood residents' participation in infrastructure planning increases the civic capacity of those residents and their communities.

PREVENTION OF HEALTH RISKS AND PROMOTION OF GOOD HEALTH

A community's infrastructure should provide healthy environments and promote healthy lifestyles for low-income communities of color. A growing body of evidence on health equity shows that people of color experience higher proportions of health risks and lower proportions of conditions that promote wellness (Bradman et al. 2005; Dunn et al. 2006; Maantay 2001; Maantay 2002). A variety of factors are related to health inequities, including:

- housing conditions and opportunities;

- levels of civic capacity and political efficacy;

- proximity to sources of toxics;

- the distribution of health care facilities;

- access to parks, recreational facilities, and healthy foods;

- the distribution of liquor stores; and

- the stresses of urban life, particularly in distressed or neglected communities.

Community infrastructure planning can play a critical, but by no means exclusive, role in preventing health risks and promoting wellness.

The disproportionate distribution of stop signs, stoplights, sidewalks, crosswalks, and similar pedestrian safety measures contributes to disproportionate numbers of pedestrian accident victims among low-income people and people of color (Harwood 2003).

Community infrastructure planning can play a critical, but by no means exclusive, role in preventing health risks and promoting wellness. First, localities and service providers should collaborate to ensure that health care services and facilities, as well as emergency response services, are not only equitably distributed but also adequate to address the increased levels of health risks that low-income people of color face. For example, cities should locate first-responder emergency personnel and equipment as close as possible to residential communities, schools, and care facilities located near potential sources of toxic releases.

Second, an aggressive local housing policy can contribute to public health and can reduce public-sector costs incurred due to the harms and risks associated with inadequate housing conditions, including health care costs, building and fire safety responses, crime and law enforcement responses, child development problems and poor educational performance resulting from conditions like lead paint exposure or frequent moves from one unstable housing situation to the next, and foregone economic activity and tax revenues. An aggressive local housing policy would improve the quality and environmental conditions of affordable housing in the area, locate housing of all types in healthy and safe locations, and enhance home ownership opportunities for low-income people of color.

Third, the siting, development, and operation of public facilities should avoid concentrating pollution-generating activities in close proximity to residences, schools, and care facilities. These pollution-generating facilities could include diesel bus depots, freeways and freeway intersections, rail yards where rail cars containing toxic chemicals sit, and sewage treatment facilities.

Fourth, sidewalks, parks, pedestrian and bicycle paths, traffic calming devices (e.g., stop signs and stoplights), open space, natural areas, and recreational facilities should be provided in low-income neighborhoods in equitable and adequate quantities to contribute to healthy conditions (Day 2006). A growing body of planning literature and practice emphasizes the importance of planning communities for physical activity and enjoyment of natural environments. Two especially valuable American Planning Association Planning Advisory Service reports on these topics are those edited by Marya Morris: 1) *Integrating Planning and Public Health: Tools and Strategies To Create Healthy Places* (Morris 2006a); and 2) *Planning Active Communities* (Morris 2006b). Planners should also heed social ecologist Stephen Kellert's book *Building for Life: Designing and Understanding the Human-Nature Connection* (2005), which shows how the human experience with natural environments is critical to human physical and mental well-being and offers planning ideas to promote healthy places. Planners and planning officials should intentionally seek the means of incorporating these principles into the planned environments of low-income communities of color.

Health impact assessments can be more effective if community residents have discussed and identified the features of healthy neighborhoods.

San Francisco Department of Public Health

Fifth, the environmental impacts of public infrastructure projects, decisions, and planning should be analyzed using emerging systematic tools for doing so. Health impact assessment (HIA) is a tool that is especially helpful for analyzing the health impacts of specific decisions and projects, similar to environmental impact assessments (Morris 2006a, 73-80). According to Morris (2006a, 74), the HIA process has five steps:

1. Screening: Identify projects or policies for which an HIA would be useful.

2. Scoping: Identify which health impacts should be included.

3. Risk assessment: Identify how many and which people may be affected and how they may be affected.

4. Report results to decision makers: Create a report suitable in length and depth for audience.

5. Evaluate impact on actual decision process.

For example, Michigan's Ingham County Health Department:

used GIS to create a Health Impact Assessment (HIA) planning matrix that planners in 78 local government units can use to assess the impact of county development projects on health. The matrix enabled planners to study the impacts in several categories, including water quality, wastewater disposal, air quality, solid and hazardous waste disposal, noise impacts, social capital, physical activity, and food systems. In terms of physical activity, questions asked include:

- Does the project provide mobility options for those who cannot drive?

- Does the project contain elements that enhance feelings of neighborhood safety?

- Does the project provide safe routes for children to walk to and from school?

- Does the project contain design elements to calm traffic?

- Does the project present unsafe conditions or deter access and free mobility for the physically handicapped?

- Does the project include pedestrian crossing signals and pedestrian refuge islands on the median? (Morris 2006a, 83)

A tool that may be more appropriate for developing a comprehensive infrastructure plan that addresses community health needs is the Protocol for Assessing Community Excellence in Environmental Health (PACE-EH), developed by the National Association of County and City Health Officials (NACCHO). NACCHO has published a *PACE-EH Guidebook* (NACCHO 2007a) and a *PACE in Practice* (NACCHO 2007b) manual that is continuously updated to provide ideas, methodologies, analytical tools, and examples to help localities develop community-based participatory processes that:

1) characterize and evaluate local environmental health conditions and concerns;

2) identify populations at risk of exposure to environmental hazards;

3) identify and collect meaningful environmental health data; and

4) set priorities for local action to address environmental health problems (NACCHO 2007a, 1).

The last task should create policy goals and priorities for health-promoting public infrastructure development.

Finally, efforts aimed at the health impacts of environmental and land-use conditions in low-income and minority neighborhoods can stimulate numerous diverse efforts to address the intersection of land use and health. For example, in the 1990s, community activists in Louisville's West End began pushing for attention to the relationships between the disparate health conditions suffered by West Louisville residents and the presence of chemical industries in West Louisville's Rubbertown area. An environmental justice group, the West Jefferson County Community Task Force, was formed as a partnership of West Louisville residents and the University of Louisville, to study the issues and facilitate community participation in addressing the West End's environmental conditions and health harms. The Task Force, with the participation of a wide range of stakeholders, was influential in the creation of an innovative local air toxics regulatory program, adopted and administered by the Louisville Metro Air Control District. However, while some of the work of the Task Force and West Louisville residents continues to focus on air toxics and air pollution controls, a variety of additional efforts have arisen, including:

- a West Louisville visioning process, designed to gather input from West Louisville residents and stakeholders about the future of the West End, including economic development, housing, and planning policies;

- a brownfields education, planning, and redevelopment multistakeholder initiative in the Park Hill Corridor of the West End, aiming to address the intersection of pollution, land-use conditions, and economic development through participatory planning organized around brownfields (organized by the Louisville Metro Development Authority, the University

A tool that may be more appropriate for developing a comprehensive infrastructure plan that addresses community health needs is the Protocol for Assessing Community Excellence in Environmental Health (PACE-EH).

of Louisville's Center for Environmental Policy and Management), and the Center for Neighborhoods, through an EPA-funded grant);

- the creation of the nation's first Center for Health Equity in the Louisville Metro Department of Public Health and Wellness and the launching of several ambitious programs to study health equity and to improve the capacity of low-income communities of color to influence policies and practices affecting their health; and

- the launching of a community-based land-use and environmental conditions assessment project, involving students from the West End's Central High School, in partnership with the University of Louisville's Center for Land Use and Environmental Responsibility and West Louisville community groups, to map and analyze various land-use and environmental conditions, incorporating health impacts and conditions into the analysis.

PUBLIC PARTICIPATION

As with every aspect of land-use planning and regulation discussed throughout this PAS Report, environmental justice means public participation—especially the full and meaningful participation of low-income and minority residents—in public infrastructure planning and development. This should begin with local neighborhood participation in an assessment of current conditions, assets, and needs. This participation should include neighborhood-based planning for the infrastructure within the neighborhood, neighborhood forums about local or regional infrastructure plans, residents' active participation in local or regional public hearings and meetings, and the representation of low-income people and people of color on regional advisory and policy-setting bodies. The community should have a substantial voice in setting public infrastructure priorities. It should also play an active and meaningful role in selecting locations, designs, and other characteristics of projects. See Chapters 3 and 5 of this PAS Report for additional details about participatory principles and methods.

CONVERSION OF UNDERPERFORMING ASSETS INTO PERFORMING ASSETS

One of the injustices that low-income and minority communities face is the disproportionately high number of "underperforming" properties in their neighborhoods. These underused properties may be vacant or abandoned, occupied by marginal or declining industries or businesses, or constrained from development by environmental contamination, poor transit, bankruptcy, absentee ownership, crime and safety concerns, or negative perceptions of the areas, including those arising from racial or ethnic prejudices.

In addition to being saddled with underperforming properties, the community may not be able to take full advantage of the opportunities made possible by its land assets, labor assets, and sociocultural assets (i.e., social capital and cultural capital), or may not be putting these assets to their highest and best use. One reason for missed opportunities and underperformance of community assets is the lack of adequate public investment and physical infrastructure. For example, exciting waterfront restoration projects or plans along the Harlem River and the Bronx River in New York City and the Anacostia River in Washington, D.C., all with the participation and/or leadership of low-income people of color, demonstrate the untapped potential of many waterfront areas in low-income and minority communities nationwide in the absence of planning and public investment.

In general, planners and local officials should work with community residents to evaluate the assets of each neighborhood or district for its underperformance and for its performance potential. Asset performance analysis, though, may have an unfortunate tendency to favor income production and private market uses.

One of the injustices that low-income and minority communities face is the disproportionately high number of "underperforming" properties in their neighborhoods.

Local officials should resist treating features of the local community merely as income-producing assets with solely economic value or basing policies solely on economic factors. Features of neighborhoods, cities, and regions have political, social, cultural, moral, and ecological value. This value cannot in many cases, and should not in most cases, be monetized (i.e., reduced to measurement solely in dollars) or commodified (i.e., treated as a good to be exchanged in private markets). Maximization of income from a neighborhood asset can in some circumstances created distributional inequities if the neighborhood residents are not receiving the benefits of that asset's use (e.g., jobs and profits that go to people who do not live in the neighborhood). Private markets can perpetuate, amplify, or even cause social, political, and economic inequity.

At the same time, however, the economic value and productivity of neighborhood assets are important to neighborhood residents, local governments and communities, and the regional economy. Market-based values of community assets are highly relevant to community planning. The critical tasks are: 1) to evaluate the productivity and potential of various community and neighborhood assets across several different values, some of which may be quantitative and some of which may be qualitative; and 2) involve neighborhood residents, local officials, and market participants in participatory discussions and decision-making processes about the highest and best uses of these assets.

All too often, environmental injustices are created or perpetuated by fragmentation of both policies and their implementation.

POLICY INTEGRATION AND COORDINATION

Incorporation of environmental justice into public infrastructure development and policy requires a coordinated and integrated approach. All too often, environmental injustices are created or perpetuated by fragmentation of both policies and their implementation.

One reason for this fragmentation is the fragmentation of staff responsibilities in local governments. Different aspects of public infrastructure might be coordinated by the city engineer, city parks department, city water department, local sewer district, transit authority, school district, and so forth. Many localities have separate public housing staff and redevelopment staff who have distinct responsibilities both from one another and from the planning staff. The lines of authority may be separate as well, not converging except at the level of the chief city administrator or the mayor and/or city council. Responsibility for local brownfields programs may fall to staff who are environmental specialists, staff who are economic development specialists, planning staff, or unfortunately in some cases to no one in particular. In addition, responsibilities for facilities and services like schools, public transportation, water facilities, and sewer systems may fall within the jurisdiction of legally and politically separate local districts or authorities.

Local officials may wish to create an environmental justice coordinating task force, composed of top officials in planning, housing, redevelopment, brownfields, transportation, and other public works divisions of the local government. Ideally, representatives from other local governmental entities, such as school districts, water districts, sewer or sanitation districts, and the like, would also participate. The role of the task force would be to coordinate infrastructure development policies to promote environmental justice principles, as well as efficiency and effectiveness in the accomplishment of policy goals.

COMMUNITY INFRASTRUCTURE

Planners and local officials must be attentive to the equitable provision, location, maintenance, usefulness, and accessibility of a wide range of community infrastructure, including the following:

- Parks

- Recreational areas and facilities

Particular attention should be given to the accessibility of neighborhood facilities, particularly by health-promoting pedestrian access.

- Tot lots
- Trails and bike and walking paths
- Open space
- Healthy and restored streams and rivers
- Waterfront and/or beach access
- Sidewalks
- Public transit facilities and services
- Streets and roadways
- Access to highways
- Stormwater management and/or drainage facilities
- Utility services and facilities, including water and sewer systems
- Community centers and neighborhood activity centers
- Schools
- Child care centers
- Cultural centers
- Science centers and nature centers
- Zoos
- Medical facilities and hospitals
- Emergency medical services
- Fire stations and services
- Police stations and services

An environmental justice strategy for community infrastructure begins with the use of GIS tools to map the locations of the above-listed facilities, analyzed against: 1) socioeconomic demographics (including race and ethnicity); 2) population densities; and 3) transportation routes, including nonautomobile transportation options (e.g., bicycle and pedestrian pathways; routes and times of public transit). Facilities serving single neighborhoods (neighborhood facilities), multiple neighborhoods (district facilities), and the entire locality or region (community facilities) should be analyzed. A thorough analysis should reveal whether the locality has distributional inequities (i.e., there are significantly fewer, poorer, or less-accessible facilities in certain areas or neighborhoods as compared to all other areas or neighborhoods within the locality or region) or inadequacies (i.e., insufficient amounts, quality, and accessibility of facilities to meet basic public needs in an area). This analysis will identify the facilities most lacking or in need, which should receive higher priority in community infrastructure planning and implementation.

Particular attention should be given to the accessibility of neighborhood facilities, particularly by health-promoting pedestrian access. California's General Plan Guidelines state:

> Public amenities can serve to anchor a neighborhood and should be centrally located. Furthermore, locating neighborhood-serving public facilities within walking distance of most residents will encourage use and provide a sense of place. A distance of a quarter to a half mile is generally considered a walkable distance. (California Governor's Office of Planning and Research 2003, 25)

Many communities are finding or will find that two particular categories of community infrastructure have been inequitably provided, with far-reaching consequences for low-income people of color and for the localities and regions in which they live. The categories are: 1) parks, recreation, open space, and natural environments; and 2) transit-related facilities.

Parks, recreation, open space, and natural environments deserve special attention because their limited availability to low-income people of color has received very little attention until recent years. Parks, recreation areas, open space, and natural environments, including healthy watersheds and plentiful tree canopy, are essential to social well-being and a good urban environment in several different respects (Garvin 2000; Garcia and Flores 2005, 145–47; Gies 2006; Tzoulas 2007; Perkins, Heynen, and Wilson 2004; Vanderwarker 2006). They contribute to public health by providing places to exercise and be physically active, and are critically important for children to develop healthy and active lifestyles, especially as obesity, inactivity, and related health problems reach alarming levels among inner-city children of color (Gies 2006, 9–10; Garcia and White, 2006, 8–9; Gordon-Larsen et al. 2006; Morris 2006b, 15–17). They contribute to mental health by reducing the stresses of the built environment and crowded, busy urban life, by facilitating the connections to nature necessary to healthy human development, and by forming enjoyable places to play and gather with others. They support programs that offer alternatives to destructive behaviors like crime, drugs, and gang activity. They reduce urban temperatures; reduce fuel usage and costs; filter pollutants from the air, water, and soils; aid ecosystem functions like flood control and pollination; and support biodiversity. They increase property values in the surrounding area and create a sense of place. They are places of cultural expression and social organization. They are democracy-enhancing places where diverse peoples can gather and interact as equals in a democratic commons, as envisioned by pioneer landscape architect Frederick Law Olmstead. The many benefits of parks have been documented in APA's City Parks Forum project, which offers helpful, free briefing papers for community use at www.planning.org/cpf/briefingpapers.htm.

Studies show that low-income people of color do not have equal access to parks, recreation, open space, and natural features, such as trees and bodies of water (e.g., streams, creeks, rivers, lakes, oceans). For example, a study in Los Angeles shows that the city has only 0.3 acres of parkland per 1,000 residents in inner-city areas, compared to 1.7 acres per 1,000 residents in areas with higher proportions of white and higher-income people, and the National Recreation and Park Association's recommended six to 10 acres of parkland per 1,000 residents (Garcia and Flores 2005, 149). A similar report on San Francisco's East Bay parkland, using a variety of statistics, also reveals a persistent pattern of unequal access to parks for minorities and low-income people (Kibel 2006). A study of 405 communities nationwide concluded that there was a 57 percent probability of a bike path being present in communities with only a 1 percent poverty rate, whereas there was only a 9 percent probably of a bike path existing in communities with a 10 percent poverty rate (Powell et al. 2004). According to another study, adolescents in high-minority, lower-educated communities had only half the probability of access to exercise facilities than adolescents in low-minority, higher-educated communities, with a substantially higher probability of teens in minority neighborhoods being overweight (Gordon-Larsen et al. 2006). Studies of the distribution of urban forests (tree canopy) and urban reforestation activities in Milwaukee show unequal distribution by race, ethnicity, and/or renter status (Heynen, Perkins, and Roy 2006; Perkins, Heynen, and Wilson 2004). Nonetheless, each locality is different. A careful study using GIS tools to measure park service areas based on pathways of access found that there were no racial, ethnic, or income disparities in Bryan, Texas (Nicholls 2001).

Planning officials and planners who are committed to equitable access to parks, recreation, open space, and natural environments in their communities can adopt an eight-step process:

1. *Commit to principles of excellence in park, recreation, open space, and natural areas planning.* The Trust for Public Land's seven factors of excellence (based on research on successful park practices in cities nationwide) offer

Parks, recreation, open space, and natural environments deserve special attention because their limited availability to low-income people of color has received very little attention until recent years.

Low-income communities of color should be connected to comprehensive regional webs that link communities, parks, schools, beaches, forests, rivers, mountains, trails, green space, transit, and other infrastructure

ideal principles to guide the assessment, planning, and implementation of an equitable parks strategy:

- A clear expression of purpose
- An ongoing planning and community involvement process
- Sufficient assets in land, staffing, and equipment to meet the system's goals
- Equitable access
- User satisfaction
- Safety from crime and physical hazards
- Benefits for the city beyond the boundaries of the parks (Harnik 2006, 11). An eighth factor of excellence that should be added is respect for the health and integrity of natural environments and ecosystem functions (Arnold 2006).

2. *Involve the community in assessment, planning, and implementation activities.* Public participation is critical. Transparency in the process is also critical.

3. *Analyze the accessibility distribution of parks, recreation, open space, and features of the natural environment (e.g., water bodies, tree canopy, native vegetation) by race, ethnicity, and income throughout the locality.* Use GIS tools and methods that identify service areas based on linear lengths of automobile and pedestrian access routes using a shortest-path-to-the-nearest-point-of-access algorithm, and compare these service areas to demographic data (Nicholls 2001). Analyses using straight-line ("as the crow flies") methods and concentric circles from centers of parks do not accurately measure true access by residents (Nicholls 2001). In addition, compare overall statistics about amount of acreage per 1,000 people for various areas of the city to the National Recreation and Park Association's recommended six to 10 acres of parkland per 1,000 residents.

4. *Consider barriers to access.* Such barriers may include: the lack of cars among minority and poor residents, who must rely on walking, biking, or public transit to reach parks and open space; the affordability of park user fees; park entrances; safety concerns; and access to people with physical disabilities.

5. *Plan for multiple facilities serving various needs.* For example, low-income and minority residents need equitable access to sufficient numbers of facilities at the levels of the neighborhood (tot lots and neighborhood parks), the area (swimming pools, plazas, greenbelts), and the region (regional parks, trail systems, beaches or waterfronts). Low-income communities of color should be connected to comprehensive regional webs that link communities, parks, schools, beaches, forests, rivers, mountains, trails, green space, transit, and other infrastructure (Garcia and White 2006, 22). Substantial investments should be made in inner-city neighborhoods and areas in urban tree canopy and reforestation, not solely on public parkland but also in other public areas (e.g., in medians, along sidewalks, in plazas and government centers) and on private property (i.e., through programs to make trees available to residents) (Heynen, Perkins, and Roy 2006; Perkins, Heynen, and Wilson 2004). Low-income and minority neighborhoods should be targets of efforts to restore watersheds and water bodies (streams, creeks, rivers, lakes, coasts) in their areas and to control runoff and water pollution, in conjunction with participatory planning for residents' access to and enjoyment of their watershed features (Vanderwarker 2006; Arnold 2006).

6. *Identify clear purposes for park, recreation, open space, and natural environment projects that have community input and support.* Also, avoid design features that will undermine the parks' purposes and vitality. Jane Jacobs cautioned against assuming that parks and open space are automatically beneficial to the community, especially where the parks: 1) lack specific beneficial purposes that are consistent with, not antithetical to, organic neighborhood vitality; 2) are not designed for public safety and safe use; or 3) essentially replace existing residents (i.e., as a blight-eliminating mechanism), instead of meeting their self-identified needs (Jacobs 1961, 116–45). In contrast, a good planning process has clear goals, participatory methods, and careful attention to neighborhood-oriented design and context.

7. *Look for unexpected sites for parks, recreation areas, open space, and restored natural environments.* Brownfields may serve as ideal sites, if cleaned up adequately. In other words, planners need to remember that while children (a population more vulnerable to environmental contaminants than the adult population) ideally should have substantial hours of exposure to park and natural environments daily, they will do things like ingest soil, roll around in grass, play at the water's edge or in the water, pick up rocks and bugs, and the like. In addition, neighborhood residents, who are concerned about the level of cleanup of their new parks or recreation areas, will likely not use these facilities regardless of what "experts" agree are the best cleanup practices or the health-risk levels of the remaining contaminants on the property. In sum, contaminated and/or developed sites are viable sources of parks and similar facilities in low-income and minority areas only if local officials are willing to invest in the processes and substantive outcomes that make these areas not just actually safe but perceived to be safe, healthy, and desired places for people to gather and play. For example, Los Angeles is converting abandoned rail yards into parks and has planned for the revitalization of the Los Angeles River after a successful environmental justice campaign to increase parks in inner-city areas (see the sidebar on Los Angeles Urban Parks and River Restoration). In St. Paul, Minnesota, a vacant shopping mall that had been built on top of a wetland was torn down and the wetland restored as part of a redevelopment plan for the economically distressed and ethnically diverse Phalen Corridor. (See the sidebar on the St. Paul Phalen Corridor under "Redevelopment" below.) Moreover, there are abundant examples of vacant inner-city lots becoming community gardens, promoting healthy interactions with the natural environment and with other community residents in the midst of the urban built environment (Bonham, Spilka, and Rastorfer, 2002).

8. *Be patient.* Initial efforts at comprehensive planning for parks, recreation facilities, open space, or ecological restoration may be met with community resident skepticism or even hostility, especially given the history of redevelopment projects and urban plans that have adversely affected low-income communities of color. These developments require planners and local officials to listen actively, to address underlying concerns, and to demonstrate that the planning process is truly a multistakeholder, participatory, deliberative process without the imposition of a preordained outcome on the community. The sidebars on the Los Angeles urban parks and the St. Paul Phalen Corridor illustrate that initial resident hostility does not undermine the overall planning process if the process is fair and participatory. In these examples, good planning outcomes were reached with the active involvement and input of many stakeholders.

Contaminated and/or developed sites are viable sources of parks and similar facilities in low-income and minority areas only if local officials are willing to invest in the processes and substantive outcomes that make these areas not just actually safe but perceived to be safe, healthy, and desired places for people to gather and play.

LOS ANGELES URBAN PARKS AND RIVER RESTORATION

Los Angeles stands out among cities nationwide for its recent efforts to address racial, ethnic, and socioeconomic inequities in parks, open space, and recreational areas. It is a city that is park-poor and that rejected a 1930 Olmstead plan for abundant parks and riparian buffers, yet is now trying to redress these planning mistakes. Environmental justice advocates focusing on these particular issues and calling themselves an "urban parks movement"—a coalition of community-based environmental and social justice groups with strong leadership by The City Project (formerly of the Center for Law in the Public Interest) —articulate a compelling vision for infrastructure equity and are enjoying success in reshaping land-use planning in the Los Angeles area. They have been joined by local and state leaders, planners, architects, and academics. Their vision integrates concepts of equal justice, vibrant democracy, public health, ecological sustainability, and intergenerational planning. Several of their greatest successes have involved the concreted Los Angeles River and adjacent or nearby lands targeted for nonpark development.

One success involves a vacant 32-acre crescent-shaped parcel of land called the Cornfield. It is the former site of a railyard near Chinatown, Union Station, downtown, Olvera Street (the original 18th century Los Angeles pueblo settlement), a major public housing project, and the Los Angeles River. Eighty-nine percent of the people who live within five miles of the Cornfield are people of color. Initially slated for a 900,000-square-foot manufacturing and warehouse development, this environmentally contaminated site with a central and historic location became the subject of intense interest by a broad coalition of local residents, environmentalists, urban planners, parks advocates, and environmental justice activists. Although community residents initially expressed skepticism and even hostility towards environmentalists' vision that the area could be a public

park, a consensus and an effective broad-based coalition soon emerg[ed] Although city officials approved the warehouse development p[ro]posal, the developer faced a litigation loss over the city's failure [to] prepare an environmental impact report for the development. The U[S] Department of Housing and Urban Development announced tha[t it] would withhold promised subsidies of $12 million until a full en[vi]ronmental impact report had been prepared. City political leaders[hip] changed with a new election, during which all the major candida[tes] announced their support for a park at the Cornfield. City officials a[nd] community advocates reached a settlement with the developer [in] which the developer would sell the site to the State of California fo[r a] state park and abandon its proposed warehouse development.

Another nearby former railyard, Taylor Yard, is also be[ing] converted into a state park, after community residents and en[vi]ronmental justice advocates were successful in opposing a propos[ed] industrial warehouse development for the site (again with litigati[on] that set aside the aside the approval for failure to prepare a full e[n]vironmental impact report). The developer sold an area known [as] Parcel D to the State of California for a 40-acre park that will be p[art] of a 130-acre park. This park will have soccer fields, a running tra[ck] natural parkland, a picnic area, bike paths, and an amphitheater[.]

These two new parks are part of a larger vision for urban par[ks,] recreational and cultural resources, and ecological restoration acti[vi]ties in the Los Angeles area. The broader plan, parts of which a[re] still in stages of proposal and advocacy, includes the creation of

The Rio de Los Angeles State Park at Taylor Yard is one of the project that arose from the Los Angeles Urban Parks and River Restoration Program. Clockwise: a bench testifying to the neighborhood's culture; an aerial view of the reconstruction area; kids enjoying a splash; and dancers in colorful dresses celebrating the park's opening.

ritage Parkscape—like the Freedom Trail in Boston – that will
k the Cornfield, Taylor Yard, and the Los Angeles River with 100
er cultural, historical, recreational, and environmental resources
he heart of Los Angeles" (Garcia and Flores 2005, 160). The plan
o includes restoration of the Los Angeles River as an ecologically
d culturally vital focal place for the entire region. Much of the
er is currently concreted and fenced, serving as a flood channel
tead of a public natural resource. Public consensus seems to be for
ne kind of "greening" of the river as part of a Los Angeles River
kway, funded in part by bonds approved by voters. In fact, this
ader "urban parks" plan includes substantial public funding by
levels of government for park acquisition and development. An
mple is California's Proposition 40, which provided $2.6 billion
parks, clean air, and clean water, and which was passed statewide
th the strong support of minority and low- to moderate-income
ers. Moreover, parks have been or are being created in minority
nmunities that are not adjacent to the Los Angeles River. These
lude a two-square-mile park in predominantly African-American
ldwin Hills, which is supported by the state-created and state-
ided Baldwin Hills Park and Conservancy, and a 140-acre park in
cot Hills, in predominantly Latino East Los Angeles.

Several lessons from the Los Angeles urban parks movement deserve
ntion. One is that the development of consensus about particular
jects and parks did not happen immediately or smoothly. Eventually
road coalition emerged, accommodating the interests of not only those

groups, like The City Project, that seek active park uses, such as soccer fields
and cultural sites, but also those groups, like Friends of the Los Angeles
River, that seek more passive uses and restoration projects that protect
the natural environment. Interestingly, the stakeholders have managed to
identify sites and parts of sites that are context-appropriate for particular
kinds of uses. For example, the proposed park in Ascot Hills has shifted
from leveling hills for soccer fields to preserving the natural landscapes
of the hills for more passive recreational uses.

A second lesson is that ultimately both developers and local officials
came to the table to resolve the land uses of particular sites in ways that
met the goals and visions of the surrounding community residents, as
well as the needs for park equity and growth. Some engineering and
planning officials have remained reluctant to think of the Los Angeles
River as anything but a flood control channel to be managed by experts,
but this resistance is being slowly worn down by a compelling alternative vision and a growing political consensus for a greening of the Los
Angeles River and its lands. This shift in policy, as slow, conflicted, and
painful as it sometimes can be, is similar to the shift in Los Angeles'
harmful appropriation and use of Mono Lake waters to practices of conservation and reclamation (Arnold 2004). As with L.A.'s consumption
of Mono Lake waters, litigation played an important role in stopping
industrial redevelopment of the Cornfield and Taylor Yard, but as a way
of facilitating collaborative problem solving and a negotiated outcome,
not as an outcome determinant (Arnold 2004).

The third lesson is that an effective environmental justice strategy
for land use requires that community residents identify their visions
and plans for lands that are the subject of development goals, not just
opposition to development proposals. The urban parks movement at-
tributes its successes to its ability to articulate a compelling vision for
sites like the Cornfield, Taylor Yard, and Baldwin Hills. ▪

Sources: The City Project 2007a; The City Project 2007b; Desfor and Keil 2004; Garcia et al.
2002; Garcia at al. 2004; Garcia and Flores 2005; Gottlieb and Azuma 2005; Kibel 2004.

Photos courtesy of The City Project

The second category of community infrastructure receiving special attention is transportation. Environmental justice issues arise in transportation infrastructure in five ways:

1. Disproportionately fewer transportation services, less access to various types of transportation, and greater barriers to mobility among low-income people of color than among the general population

2. Disproportionate location of environmentally burdensome transportation facilities, such as diesel bus depots, freeways and freeway interchanges, and freight rail lines and yards in or near low-income communities of color

3. Poor maintenance of streets, sidewalks, and bike paths in low-income communities of color

4. Inadequate emergency evacuation plans for low-income people of color, as the Hurricane Katrina debacle in New Orleans illustrate

5. Underrepresentation of low-income people and people of color on local and regional transportation planning bodies, as well as barriers to the participation of low-income people of color in transportation policy and planning (AASHTO 2006)

While a comprehensive environmental justice analysis of transportation policy is beyond the scope of this report, two particular observations merit attention. First, environmental justice advocates have gathered many examples and studies of the burdens that low-income communities of color face from transportation-related air pollutants, noise, traffic congestion, exposure to toxic chemical spills in association with these chemicals' transportation, disruption or even destruction of neighborhoods (or parts of neighborhoods) from the construction or expansion of freeways and other transportation facilities (Bullard and Johnson 1997; Bullard, Johnson, and Torres 2000; Pulido 2000; Maantay 2001; Bullard, Johnson, and Torres 2002; Jakowitsch 2002; Maantay 2002; Bullard 2005; AASHTO 2006). Second, consider a study of board membership for 50 large metropolitan planning organizations (MPOs) that plan regional transportation facilities and services receiving federal and state transportation funds (Sanchez and Wolf 2005). The study showed that only 12 percent of voting members of MPO boards are members of racial or ethnic minorities, even though the racial and ethnic minority proportion of the populations served by these MPOs is 39 percent. Moreover, 13 of the 50 MPO boards had no people of color at all.

Incorporation of environmental justice principles into transportation planning requires assessment of the distributional patterns of the following five categories of transportation policy and changes in transportation conditions to remedy or prevent disparities and burdens in low-income communities of color in each of these five categories:

1. Transportation services

2. Externalities of transportation infrastructure (i.e., pollution, noise, neighborhood disruption)

3. Maintenance of transportation infrastructure, including pedestrian-supporting infrastructure

4. Emergency evacuation plans, especially for those with the least control over their own evacuation options

5. Representation and participation of diverse populations in transportation policy and planning

A study of 50 MPOs showed that only 12 percent of voting members of MPO boards are members of racial or ethnic minorities, even though the racial and ethnic minority proportion of the populations served by these MPOs is 39 percent. Moreover, 13 of the 50 MPO boards had no people of color at all.

The following specific policy recommendations are adapted for local and regional planning activities from federal transportation policy recommendations offered by Nancy Jakowitsch (2002, 7–9) of the Surface Transportation Policy Project:

1. Require accountability for equity outcomes from local and regional transportation agencies. Evaluate long-range transportation plans, transportation improvement programs, and environmental reviews of specific transportation projects to determine whether they actually advance environmental justice principles.

2. Build capacity for equitable transportation planning.

3. Establish clear policy guidance, goals, and performance measures for equity outcomes in the transportation planning process that can be customized with community input.

4. Use modern modeling tools and planning techniques, including consideration of cumulative impacts of transportation investments, alternatives, and the strongest data and science to evaluate impacts. Improve and mainstream the use of models to predict impacts on different socioeconomic groups.

5. Present geographic data visually to depict long-range outcomes related to both transportation and land-use scenarios.

6. Improve opportunities for public participation. Evaluate public involvement processes to determine and remove participation barriers facing minority and low-income populations in transportation decision making. Make resources available to community groups, nonprofit organizations, and academic centers to actively participate, including funds for neighborhood planning grants, data collection and analysis, and community-based training. Provide information in order to ensure a more informed decision making process.

7. Collect and analyze data relating to transportation needs of different population groups. Develop and refine models of residential, employment, and transportation patterns of various low-income and minority populations. Incorporate into transportation planning the use of geographic data from multiple sources, including various government agencies. Analyze data to determine whether the benefits and burdens of transportation investments are distributed fairly.

8. Seek and facilitate effective interagency coordination during the transportation planning process, especially to address spatial and temporal gaps in transit service for low-income and minority populations.

9. Invest in intermodal transportation and choices for communities, especially economically disadvantaged communities.

10. Ensure that investments in transit facilities, services, maintenance, and vehicle replacement provide equitable benefits to communities of color and low-income communities, including use of creative ways to maximize transit benefits to these communities.

11. Use the Job Access and Reverse Commute (JARC) program to increase the frequency of existing transit service, improve evening and weekend hours, make better connections to key destinations within existing communities, and increase community and stakeholder involvement.

12. Prioritize funding for the Clean Fuel Bus Program to reduce diesel exhaust of buses in low-income and minority communities.

13. Invest in safe routes to schools to address the disproportionately high rate of pedestrian fatalities among low-income children of color.

14. Modernize and improve transit services. Use Intelligent Transportation Systems that can inform the traveling public (e.g., provide information about when the next bus will arrive; use GIS-mapping programs to identify low-income neighborhoods, employment centers, childcare facilities, and other route- and scheduling-related data; and provide buses with improved access to people with disabilities).

15. Remove barriers to transit-oriented development. Plan and create incentives for mixed-use, mixed-income developments that create places as well as transportation nodes for intermodal transfers. Develop these TODs without displacement of existing residents.

16. Plan and create incentives for low-income housing near transit centers.

Planning for equitable housing infrastructure requires a comprehensive, multifaceted approach that avoids overemphasizing the "housing policy du jour."

HOUSING

Limited housing opportunities, unaffordable housing markets, and poor physical conditions are some of the most far-reaching land-use conditions that low-income communities and communities of color face, exacerbating concentrations of poverty, health inequities, barriers to socioeconomic mobility, and exposure to harmful environmental conditions (Barnett 2003, 63–75; Bradman et al. 2005; Crowley 2003; de Souza Briggs 2005; Dunn et al. 2006; Krieger and Higgins 2002; Maantay 2001; Schill and Wachter 1995; Squires 2007). An environmental justice policy for housing seeks to secure in a given community an adequate supply of affordable, safe, good-quality housing accessible to low-income people and people of color.

While the goal is simple, the task of achieving it is extraordinarily complex. This is due to: 1) the wide variety of housing needs and issues within a community; and 2) the number, complexity, and intractability of barriers to achieving needed housing supplies (Crowley 2003; de Souza Briggs 2005; Goodno 2002; Press 2007; Schwartz 2006; Smith and Furuseth 2004; Talen 2006). Far too often, a one-size-fits-all housing policy fails to address the multifaceted dynamics of housing supply and demand. Housing inequity results from both problem complexity and policy simplicity.

Planning for equitable housing infrastructure requires a comprehensive, multifaceted approach that avoids overemphasizing the "housing policy du jour," whether new urbanist infill, market-based vouchers, mixed-income housing, urban renewal, mixed-use workforce housing, or other such models.

Some low-income people of color seek or accept geographic mobility, whereas others insist on remaining in existing neighborhoods where they have social networks and cultural history. Some seek affordable home ownership opportunities, others seek affordable rental opportunities, and still others seek rehabilitation of existing housing in disrepair. Low-income people of color who live in central cities are likely to face different housing issues than are low-income people of color who live in suburbs or rural areas, but poverty and lack of affordable housing exist in all of these settings. Some low-income people of color, especially the elderly and those with disabilities, may require certain services to accompany their housing options (Libson 2005-2006). Standard assumptions about proximity of housing to the locations of jobs and the transportation options between home and work may not hold true for certain subpopulations of a local community, including some groups of low-income people and some groups of racial and ethnic minorities (Smith and Furuseth 2004; Press 2007; Weis 2007). Some homeowners in low-income communities of color seek to improve their condition through market-based wealth created by rising property values (i.e., increasing equity in one's home), whereas other residents of those same communities fear that rising property values will displace them and put housing beyond their capacity to afford it. While these observations may be intuitive to most planners, the obstacles to achieving housing plans and policies that address all of these needs often result in inadequate housing for low-income people of color.

Several principles characterize an environmentally just and socially equitable approach to planning for local housing needs. First, planners should gather as much differentiated, nuanced data about the locality's various types of housing needs as they possibly can. General aggregate data about the region's overall need for affordable housing are not detailed or differentiated enough to give an accurate picture of a diverse set of housing issues and needs.

New Columbia, a Hope VI redevelopment project in Portland, Oregon, integrates affordable housing development with creation of healthy, livable, vibrant places of community.

Photos by Mike Wert, Housing Authority of Portland

At their worst, current trends in housing polices could be used to uproot families from their neighborhoods and paternalistically place these families in areas where they are isolated and keenly perceive their "differentness" from higher-income, nonminority neighbors.

Second, affordable housing policies should balance mobility-oriented plans with rootedness-oriented plans. Current models of housing policy, both nationally and in many cities, favor the creation of new affordable housing units dispersed throughout the locality or region in mixed-income neighborhoods (Barnett 2003, 63-75; Berube and Katz 2005; Thomas-Houston and Schuller 2006; HUD 2004; Weis 2007) . These mobility-oriented models recognize the growth of job opportunities in suburban and exurban areas, as well as the presence of low-income people in these communities. They attempt to avoid the various social, economic, health, and safety problems associated with the concentration of poverty in central-city neighborhoods. Indeed, programs like Hope VI aim to deconcentrate poverty by replacing high-density public housing projects with dispersed, lower-density, mixed-income units. Contemporary housing policies are based on research showing economic and social mobility for low-income people who reside in mixed-income neighborhoods (Boston 2005). They attempt to prevent or reverse racial and ethnic segregation within metropolitan areas.

One manifestation of mobility-oriented models of affordable housing is the public policy favoring increased home ownership opportunities for low- and moderate-income people because of the sociopsychological benefits of home ownership, the wealth increase from asset appreciation, and the greater acceptability of affordable owner-occupied homes in mixed-income neighborhoods in comparison to rental housing units (Retsinas and Belsky 2002; Rohe and Watson 2007).

Mobility-oriented models of affordable housing supply have drawbacks and inadequacies, though (Thomas-Houston and Schuller 2006; de Souza Briggs 2005; Crowley 2003; Talen 2006; Weis 2007). They fail to meet needs for affordable housing in existing neighborhoods of low-income people of color characterized by healthy social networks, cultural traditions and identities, support systems, and proximity to jobs held by existing residents. For example, a recent study of single mothers in Louisville, Kentucky, shows that they "are able to benefit from living in the city, because of the proximity of well-paying jobs in the health care industry, job search and job training centers, more reliable and frequent public transportation, dozens of state-approved child care facilities, and social services. It was also found that single mothers living in Louisville's inner city benefit from living near other poor, single mothers, for the help and support that these neighborhood networks can provide" (Weis 2007).

At their worst, current trends in housing polices could be used to uproot families from their neighborhoods and paternalistically place these families in areas where they are isolated and keenly perceive their "differentness" from higher-income, nonminority neighbors. Dispersion models may become tools—used either purposefully or inadvertently—to disperse the political power of racial and ethnic minorities within a locality. Policies of developing new housing stock in mixed-income communities may fail to invest sufficient resources in rehabilitation of existing affordable housing stocks that have simply aged or fallen into disrepair in low-income neighborhoods of color. Dispersion models may take attention away from the need to enhance the infrastructure of existing low-income neighborhoods of color, including schools, public transportation, parks, and economic/work opportunities. While those low-income people of color who move to new affordable housing in the suburbs might have access to better public infrastructure and services, those who remain in underserved central cities would continue to experience increasing disinvestment and substandard conditions.

The challenge for housing and planning officials, therefore, is to make substantial investments in both: 1) new housing opportunities in mixed-income neighborhoods throughout the metropolitan area (i.e., a dispersion-oriented model), and 2) new or improved housing opportunities and other neighborhood infrastructure in existing low-income neighborhoods of color (i.e., a rootedness-oriented model). A mix or balance of both approaches is needed.

Third, placing housing in or near industrial areas typically violates principles of justice, public health, and good planning, but may be needed or even appropriate under certain circumstances. As discussed throughout this PAS Report, residential land uses are incompatible with industrial and other pollution-generating land uses. However, a simplistic policy of not placing any new housing within an established distance from existing industrial facilities or land zoned for industrial use might violate other environmental justice principles. These principles might include community empowerment and neighborhood residents' control over their environment, equitable provision of affordable housing options for low-income people of color, and strengthening of low-income neighborhoods of color through improvement of the physical, social, cultural, and political environment. Context makes all the difference. For example, development of affordable housing in a mixed residential-industrial area might be appropriate if industrial land uses are on the decline, being converted to other uses, and subject to rezoning, overlay zones, performance standards, or other controls limiting their potential impact on nearby residents. In fact, public investment in housing in such a neighborhood might be critical to revitalizing the neighborhood, strengthening the community's existing social and cultural fabric, and improving the physical environment for existing residents.

Fourth, affordable housing development policies should be balanced with economic development policies. Of course, planners' concerns with the jobs-housing balance in areas of scarce housing supply and rising housing costs are likely to necessitate planning for as much affordable housing as possible so that economic activity does not decline due to business's inability to attract or retain workers who cannot afford or find housing. On the other hand, though, both fast-growing areas and economically distressed or challenged areas require attention to facilitating business development and economic investment that will provide jobs and income to low-income people of color. The ability of low-income people of color to afford housing and the ability of the locality or region to maintain a sustainable jobs-housing balance in the long run requires the use of land—at least some land—in or near low-income communities of color for employment-producing, tax-paying, income-generating businesses. Policies supporting and facilitating microenterprise development and growth among low- and moderate-income people of color are similarly related to a balanced housing policy. Planners should explore opportunities for mixed-use developments that combine affordable housing (including home ownership opportunities) with small businesses owned and operated by area residents.

Fifth, community development in general, including redevelopment, housing development, and economic development, should contain cost-control mechanisms to prevent low-income people from being priced out of their homes and neighborhoods. Gentrification of low-income neighborhoods of color far too often characterizes redevelopment activity, including housing redevelopment. While contemporary housing programs, such as Hope VI, are expected to minimized displacement and provide relocation that meets the housing needs of relocated residents, implementation of these requirements remains flawed and some degree of gentrification can still occur. Gentrification can be merely a form of asset appreciation and community revitalization, which are critically important goals. However, these goals should not result in the poorest people simply being bumped from neighborhood to neighborhood as each becomes revitalized and as housing costs (and other

Affordable housing development policies should be balanced with economic development policies.

costs of living) in the revitalized areas increase. Community land trusts are promising tools for retaining affordability and preventing displacement of low-income people of color, if used properly in partnership with the affected residents. See the sidebar on Lexington's use of community land trusts.

Sixth, planning officials should analyze local zoning codes and development standards for exclusionary zoning techniques that effectively preclude affordable housing and should reform zoning codes where these techniques exist. In addition, planning officials should watch land-use and development activities carefully for patterns of racial and ethnic residential segregation.

Arbolera de Vida declares that the Sawmill Community Land Trust is the reason she could afford to own a home in Albuquerque's North Valley.

Dory Wegrzyn

Finally, planning officials should plan for infrastructure to accompany the development of affordable housing. Equitable housing opportunities for low-income residents include transit options; sidewalks and pedestrian safety features; tot lots and open space; sewer and stormwater management facilities; accessible libraries and community centers; and similar infrastructure common to higher-income neighborhoods.

REDEVELOPMENT

The American Planning Association's Policy Guide on Public Redevelopment (2004) describes the intersection between public redevelopment programs and environmental justice principles:

> The redirection of growth into the nation's central cities, urbanized areas, inner suburbs, and other areas already served by infrastructure and supported by urban services is an essential element of the American Planning Association's vision. Local government redevelopment programs [defined as "public actions that are undertaken to stimulate activity when the private market is not providing sufficient capital and economic activity to achieve the desired level of improvement"] provide critical tools for accomplishing this goal. . . .
>
> Redevelopment activities can be used to create or leverage better housing choices, better access to goods and services and employment opportunities. This would most appropriately be done in conjunction with an overall community strategy that matches services such as job training programs with physical improvements. Communities may also choose to use redevelopment tools to create choices in underserved communities where blighting

THE LEXINGTON COMMUNITY LAND TRUST

A variety of public projects can displace low-income people of color as a result of the project's direct acquisition and use of land and as a result of the project's indirect gentrifying effects. Highway expansion and extension projects in or near low-income communities are among the projects that typically have these impacts. Transportation planners in Lexington, Kentucky, have proposed using a community land trust to mitigate these impacts from a planned highway extension.

The Newton Pike Extension will connect Interstate Highways 64 and 75 to Lexington's downtown and the University of Kentucky, easing traffic congestion. The extension will run through Southend Park (also known as Davistown or Davis Bottoms), a mixed-race neighborhood characterized by very-low-income households and substandard housing. The primary threats posed by the extension are increased development pressures from downtown and the university, which will drive up housing costs and displace residents. For 35 years, the future possibility of the extension created uncertainties that deterred redevelopment plans and the downzoning of nonindustrial properties to more appropriate, less-intensive designations. Finally, in 2003, the Lexington-Fayette Urban County Government's Planning Commission approved a Southend Park Plan for a mix of housing types and tenures, open space and parks, and commercial and institutional uses. The redevelopment plan was to advance a primary policy of preserving the existing residential occupation of the area.

Identifying development pressures, displacement, and increased housing costs as the greatest threats to Southend Park, both the Southend Park Plan and the Newton Extension Corridor Plan proposed a Lexington Community Land Trust (LCLT) as an environmental mitigation component of the extension and the accompanying redevelopment of the area. Under the LCLT, the land in the redevelopment area will be owned and managed by the trust, governed by a board consisting of one-third "lessee" representatives from the area, one-third "general" representatives of the Fayette County community, and one-third public representatives (local, state, and federal officials).

By replacing individual ownership with trust ownership, the land can be leased to area residents at affordable rents or can remain owner-occupied but subject to limits on resale and capital gains at resale, thus keeping prices low for future buyers. In other words, the land trust's ownership of the land removes it from market pressures, thus maintaining market affordability. The area subject to the LCLT's ownership and control will initially be limited to Southend Park but after 10 years could include any area within Fayette County where affordable housing is an issue. The LCLT's priorities for occupants of its housing are: 1) current residents of Southend Park; 2) low-income residents of the broader area subject to the Newton Pike Extension Corridor Plan; 3) low-income former Southend Park residents and relatives of current residents; 4) low-income residents of Fayette County; and 5) other low-income households.

The LCLT illustrates a useful tool to prevent gentrification associated with the redevelopment or revitalization of low-income and minority communities. However, the LCLT also illustrates three mistakes to avoid when considering community land trusts. The first mistake is top-down planning as the source of the community land trust. Local residents have been suspicious of the LCLT because it resulted from a government-driven, instead of community-driven, process. While leadership from government officials and planners may be strongly needed, especially in sparsely populated and economically and physically stressed areas like Southend Park, it must be more about generating a long-range process of community participatory planning and generating ideas for local residents to consider than about pushing a plan or set of ideas onto the community for its acquiescence. The second mistake is to create a trust structure with a relatively low percentage of neighborhood residents serving on the board. A sizeable nonresident presence on the board may result in conflicts when market pressures undermine housing affordability, promote a desire to sell land and property, and, consequently, undermine the trust's core purpose. Land trusts should be created not only with a greater percentage of neighborhood residents on the board, but also with a super-majority vote requirement to approve land sales (e.g., two-thirds of all board members). The third mistake is to underestimate the value of individual land ownership, especially among African-American and low-income communities that have found land and home ownership historically to be so elusive. An effective plan requires community consensus that communal ownership through a land trust is preferable to individual ownership under the circumstances. In the case of the LCLT, neighborhood residents' resistance to trust ownership has been exacerbated by the first two mistakes: top-down planning and limited resident control of the trust. ▪

Source: Bourassa 2006

Economic development and urban redevelopment efforts should include "resident ownership mechanisms" by which residents in low-income communities become asset owners, not merely political stakeholders, in new development.

conditions have not yet taken hold in order to achieve a better balance of access and choice throughout the area. In all cases, planners need to guard against redevelopment activities that are not respectful of a community's existing societal and cultural fabric.

Redevelopment can provide an opportunity to redress issues of environmental justice. Without adequate assessment of environmental impacts, however, redevelopment may have disproportionately adverse effects on the lower-income households that reside in or near redevelopment project areas. . . . Communities committed to achieving environmental justice [should] plan the location and design of public improvements and projects that are likely to have adverse environmental and aesthetic impacts – such as highways, industrial plants, and correction facilities – to ensure that they do not disparately affect areas with a substantial number of disadvantaged households.

In addition to the general principles elucidated in the APA Policy Guide, recent research has illustrated some additional specific principles of equitable redevelopment practices. One is that data-driven participatory neighborhood revitalization strategies that have widespread community support and that make public investments in distressed neighborhoods can substantially leverage private investment in those neighborhoods even to the point of being self-financing over a 20-year horizon (Galster, Tatian, and Accordino 2006).

Another point for consideration is that, despite the tendency for new urbanist projects to be located on previously undeveloped lands (greenfields), there is evidence that new urbanist inner-city redevelopment can be successful using community-based (i.e., neighborhood-driven) processes (Deitrick and Ellis 2004).

Third, planners and local officials might consider using structured groups or entities of neighborhood residents, property and business owners, and other stakeholders to evaluate the impacts of proposed redevelopment and other infrastructure projects on the neighborhood. For example, a recently enacted statute in the State of Washington authorizes the creation of Community Preservation and Development Authorities (proposed by local residents and property and business owners, but subject to approval by the state legislature) to which local and state officials may refer major public facilities, public works projects, or capital projects to evaluate their impacts (including cumulative impacts) on the cultural and historical identity of the community (Washington Substitute Senate Bill 6156, Chapter 501, Law of 2007, 60th Legislature, 2007 Regular Session, Approved May 15, 2007, except for partial veto).

Fourth, economic development efforts targeted to the expansion of minority-owned businesses, especially as a part of place-based inner-city redevelopment, are critical to maximize job opportunities for minority residents, who are more likely to be hired by minority-owned businesses than other businesses (Bates 2006).

Finally, economic development and urban redevelopment efforts should include "resident ownership mechanisms" by which residents in low-income communities become asset owners, not merely political stakeholders, in new development (McCulloch and Robinson 2001). These mechanisms include resident shareholder or partner status in projects undertaken by and with Community Development Corporations, the creation of community-building individual development accounts, shared-resident equity in business development, cooperative forms of ownership (including employee ownership), community development credit unions, community land trusts, real estate investment trusts, and others (McCulloch and Robinson 2001).

BROWNFIELDS

Brownfields are underused or abandoned sites, usually industrial, that face barriers to redevelopment or full use by actual or perceived environmental contamination. While the potentially large liabilities for hazardous contamination cleanup that accompany site ownership or operation inhibit development or use under the Comprehensive Environmental Response, Compensation, and Liability Act (CERCLA, commonly referred to as the Superfund law), a wide range of other factors also inhibit cleanups and productive uses of these sites (Davis 2002; Heberle 2006; Howland 2003; Gerrard 2001; Solitare and Greenberg 2002). These factors include:

Abandoned and/or deteriorating properties may pose risks to area residents from trash, vermin, falling structures, fires, and attraction of crime, and are visual, economic, and sociopsychological blights on the surrounding area.

- psychological perceptions of risk;

- stigma;

- uncertainties about contamination levels and the degree of cleanup needed for various kinds of future uses;

- perceptions about the safety and desirability of the areas surrounding brownfields;

- complicated legal arrangements concerning various cleanup and reuse;

- participants' responsibilities and liabilities;

- costly and time-consuming legal processes;

- often inadequate mechanisms for area resident participation in decisions about future uses and levels of cleanup;

- lack of comprehensive planning for areas in which substantial numbers of brownfields are located; and

- lack of coordination (and perhaps even cooperation) among local, state, and federal officials responsible for environmental regulation, cleanup standards, redevelopment, site management and marketing, housing, economic development, land-use planning, and other areas critical to successful brownfields redevelopment.

Brownfields pose several land-use planning challenges with a distinct "environmental justice" nature (Gerrard 2001; Solitare and Greenberg 2002; Felten 2006). First, a high proportion of brownfields exist in areas where low-income people of color live. This is not surprising given the history of mixed industrial and residential zoning in inner cities and other areas where low-income people of color reside. Brownfields impose a variety of burdens on these communities. The environmental contamination that characterizes the sites may expose area residents to heightened risks of human health hazards. Migration of contaminants into groundwater and even surface water in the area is one such concern. Even if the contamination is contained to the site or within certain structures or contained environments, the sites may be attractive nuisances to area children who become exposed either to the site-contained toxics or other hazards when they visit the sites. Abandoned and/or deteriorating properties may pose risks to area residents from trash, vermin, falling structures, fires, and attraction of crime. They are visual, economic, and sociopsychological blights on the surrounding area. They inhibit the thriving vibrancy of low-income neighborhoods of color. Moreover, they consume land that could be put to productive uses, contributing to the area economy and community sense of place, as well as to local tax revenues. Brownfields represent missed opportunities for jobs, investment, and growth in low-income communities of color. Thus, a strategy of brownfields cleanup and re-use is an environmental justice goal.

There is growing evidence, though, that neighborhood-based groups of low-income people of color do not want to deter cleanups and brownfields redevelopment with cleanup standards disproportionate to the planned land use and potential exposure of area residents to any residual contamination.

A second environmental justice issue, though, centers on the level of cleanup needed before brownfields are redeveloped or put to new uses. Low-income people of color may fear that the cleanup levels in their communities will be inadequate to eliminate health and safety risks associated with residual contamination. They may fear that risk assessments underestimate risk, exposure, and worst-case scenarios, or set the level of risk that area residents should accept at a level that is too high. They may fear that they are not receiving reliable information or that a second-class (or even third- or fourth-class) cleanup is being pushed on their community in order to accommodate businesses and political interests on the assumption that low-income people of color will not know the difference or will be willing to accept any improvement over currently unacceptable conditions. They fear that required cleanup standards will not be enforced and that agreed-upon procedures and plans will not be followed. Their belief that businesses and local leaders will expose them to unhealthy levels of contamination if financially and politically expedient to do so is built on their experiences with decades of environmental and land-use policies that place low-income people of color and sources of environmental harms in proximity to one another. They may question whether the type of new use is appropriate for a former brownfield, especially if the new use is housing or a school. These concerns cannot be dismissed or ignored; the perceptions of area residents have a very real effect on brownfields redevelopment.

On the other hand, government officials and current and future owners of brownfields may be overly inhibited by fears that low-income people of color will demand unreasonably stringent levels of cleanup. They may perceive that every brownfields cleanup in areas with substantial numbers of low-income or minority residents has considerable potential to become an environmental justice battle.

There is growing evidence, though, that neighborhood-based groups of low-income people of color do not want to deter cleanups and brownfields redevelopment with cleanup standards disproportionate to the planned land use and potential exposure of area residents to any residual contamination. Instead, these groups seek to be involved early and fully in planning, impact assessment, and approval processes. They seek to ask tough questions about risk analyses, cleanup methods, proposed land uses, and health-protecting safeguards. They demand participatory and deliberative processes that may require accommodation from other participants who are accustomed to quicker and simpler management-based or transactional decision making. Low-income and minority area residents may approach brownfields redevelopment with caution, but in most cases they are committed to achieving viable cleanups and reuses of these contaminated or potentially contaminated sites.

A third environmental justice issue relates to the type of future land use proposed for brownfields sites. Industrial reuse may be tempting. Many brownfields sites contain structures and infrastructure for industrial use. Moreover, industrial reuse typically requires lower levels of cleanup, which are cheaper and quicker than cleanups for sites with greater levels of human exposure or greater margins of safety. However, an environmental justice strategy of land use typically focuses on reducing and preventing neighborhood residents' exposure to industrial land uses. Replacing a vacant industrial site near low-income and minority neighborhoods with a new industrial facility may be inconsistent with environmental justice planning goals. Moreover, neighborhood-based planning may have identified other desired types of land uses in the area, such as affordable housing, retail businesses, health care facilities, new schools, and parks and open space. Nonetheless, the economic benefits of some types of industrial and commercial land uses may be high priorities to low-income communities of color. However, even when new industry is desired or at least acceptable, community residents have questions about how

THE ST. PAUL, MINNESOTA, PHALEN CORRIDOR INITIATIVE

The Phalen Corridor Initiative is a major, award-winning redevelopment project on St. Paul's East Side, combining environmental justice, economic opportunity, and public and private investment. A partnership of diverse community residents and neighborhood groups, area businesses, nonprofit community service providers, and government officials has been at work since 1994 to produce a comprehensive and sustainable redevelopment plan for an economically and environmentally distressed area.

The starting point of the Phalen Corridor partnership was a set of community conditions—both serious problems and valuable assets. The problems were brownfields; environmental contamination in the area; high levels of poverty; unemployment; industrial decline; poor quality and deteriorating housing; and a vacant shopping center developed over a former wetland. The assets were large parcels of underused land proximate to downtown; many good-quality, affordable homes in the neighborhood; racial and ethnic diversity; strong churches and neighborhood groups; the engagement of community residents; the involvement of major businesses like 3M, Wells Fargo, and Xcel Energy; and the commitment of government officials and staff at all levels. The presence of railroads and industry served as both problems and assets. The Phalen Corridor area had been the destination of a diverse group of immigrants from Scandinavians and Central and Southern Europeans to Africans to Latin-Americans to Asians, seeking housing close to industrial jobs. However, industry in the area declined, leaving high rates of unemployment, poverty, crime, deteriorated neighborhood conditions, environmental pollutants, and brownfields.

The critical element to progress was the collaboration of a diverse and large set of interests, organized as the Phalen Corridor—an unincorporated partnership of public, private for-profit, and nonprofit participants. Some came to the effort out of social justice concerns, while others came because of economic development goals. Over 60 partners have remained committed to the endeavor. Community workshops involved local residents and businesses in planning for brownfields cleanup and land-use visioning for the area. The process resulted in a consensus of three primary goals: a transportation corridor, job creation, and brownfield remediation. The process has had controversy and conflict at times, such as when an incorrect map for a planned industrial park-housing development showed the demolition of 350 existing homes in the Vento Neighborhood, prompting angry reaction by more than 200 residents. However, redevelopment proponents expanded the planning process with regular community meetings. At these meetings, area residents had constructive input into developing plans that achieved widespread support. For example, in the Vento Neighborhood controversy, the plan was redesigned so that only 11 existing homes would be demolished.

Moreover, the revitalization of the Phalen Corridor has been the result of many different, yet planned and coordinated, projects:

- *Phalen Boulevard, a new road that forms the basis of a multimodal transportation network in the Phalen Corridor*. The network includes existing rail lines, a bike and pedestrian trail connected to the Minneapolis-St. Paul networks of bike trails, and a bus rapid transit line. A former salvage yard and waste transfer station is being turned into a new transit facility that will create 300 jobs.

- *The reclamation of a historic wetland, the Ames Lake Wetland, from a mostly vacant and partially flooded shopping center and parking lot that had been built on top of the wetland*. The wetland serves as a central component of an urban village redevelopment concept, as well as an ecological restoration plan with replanting of indigenous species. The recreated ecosystem mimics the original natural system and improves biodiversity, yet has not achieved full wildlife habitat connectivity potential due to the development of homes on cul-de-sacs around the wetland. The wetland serves as a central environmental amenity of revitalized housing development in the area.

- *The Williams Hill Business Center, built under the authority of the St .Paul Port Authority on a formerly heavily polluted site*. This project required extensive environmental cleanup to turn the brownfield into a center attracting nine businesses and more than 400 new jobs. This is just one example of the cleanup and re-use of almost 125 acres of brownfields in the Phalen Corridor.

- *Westminster Junction, a business center using sustainable building techniques*. The center has commitments to full occupation, with a projected job base of more than 1,000 employees.

- *The use of brownfields redevelopment incentives*. These incentives include government subsidies and $1 sales of municipally owned brownfields, but, importantly, the incentives are linked to conditions related to minimum site development, the creation of one new job for each 1,000 square feet of building space, and the commitment to use local residents to fill at least 70 percent of new jobs on the site. Job retention is also a major goal of the Phalen Corridor Initiative, with a focus on revitalizing the area to make it attractive for businesses to remain.

- *The reuse of an abandoned high school*. The high school will become a new elementary school and YMCA.

- *New housing in the Phalen Corridor*. Much of this housing aims at urban densities. One major project is Phalen Crossing, a mixed-income, mixed-product development within the boundaries of the wetlands park, consisting of townhouses, condominiums, and single-family homes. Other developments consist primarily of affordable housing, including the Native American Elders Lodge, the Realife Cooperative Senior Housing, Roosevelt Homes neighborhood renovation/rehabilitation, Ames Lake and Rose Hill public housing complexes, and Habitat for Humanity housing. Some housing projects are using sustainable development techniques.

- *New retail and professional businesses*. These developments will meet the demands of new and revitalized business and residential activity, as well as to stimulate such activity.

- *Funding from many sources*. These sources include state and local funds, local community development corporations, an Enterprise Community Grant, Economic Development Administration Grant, HUD 108 loan, EPA Pilot Grant, HUD Brownfield Economic Development Initiative Grant, US Department of Transporation, grants from private foundations, such as the 3M Foundation, and private investment.

Sources: Brown 2006; Dowdell, Fraker, and Nassauer 2005; EPA 2002; Jossi 1997; Milburn 2005a; Milburn 2005b; Milburn 2006; Neighborhood Environmental Conference 2002; Peiken 2005; The Phalen Corridor 2005; www.phalencorridor.org.

There are no easy formulas for addressing future land uses of brownfields. Resolution of the issues will likely depend on the specific site, the proposed land use, and the particular set of community dynamics and goals.

many jobs will go to area residents (including job training and placement programs aimed at the facility's neighbors) and how much pollution and other community impacts will result from the facility.

There are no easy formulas for addressing future land uses of brownfields (Howland 2003; Heberle 2006; Felten 2006; Davis 2002). The resolution of the issues will likely depend on the specific site, the proposed land use, and the particular set of community dynamics and goals. However, the problems are likely to be most difficult to resolve when site-specific brownfields redevelopment activities and community-oriented land-use planning are distinctly separate functions of local government, often uncoordinated due to turf jealousies, different conceptual approaches (ways of framing the issues), resource limitations, and lack of "best practices" information about the integration of land use planning and brownfields redevelopment.

Despite the environmental justice issues that must be addressed for any brownfield redevelopment, the very existence of brownfields in low-income communities of color is an environmental justice and land-use planning challenge that can lead to comprehensive redevelopment plans and good planning for revitalized neighborhoods, achieved with the input and participation of a diverse set of local stakeholders.

The Phalen Corridor in St. Paul, described in the sidebar, is an example of community redevelopment with a strong foundation in both environmental justice principles and commitments to resolve brownfields problems. Please note that APA is currently in the midst of a research project about creating community-based brownfields strategies. Details about the project, which will culminate in a PAS Report and already has an extensive list of resources, can be found at www.planning.org/brownfields/index.htm.

To take advantage of opportunities for brownfields redevelopment that honor the principles of environmental justice, local officials should adopt several planning policies:

- Establish a primary policy for brownfields redevelopment to eliminate health and safety risks to people, including disproportionate exposure to potential harms by race and class, and to promote good health and safe environments for all peoples.

- Integrate brownfields redevelopment and area-based planning processes that seek to address environmental justice and land-use inequity issues. Identify the various types of land uses that may be appropriate for different categories of brownfields, classified by site characteristics, contamination types and levels, site locational context, available infrastructure, market-related investment and development opportunities, and community/neighborhood goals.

- Consider using brownfields redevelopment needs as a mechanism for starting neighborhood-based planning processes in low-income communities of color.

- Achieve coordination among local staff with responsibility for brownfields, economic development, housing, and land-use planning; bridge the divide between market- and site-oriented management of sites and public- and area-oriented planning of communities.

- Create participatory processes with early and full area-resident involvement in decisions ranging from brownfields planning goals to specific site cleanup and re-use plans; do not limit these processes to inform-and-consent sessions, but instead seek multiparticipant dialogue and problem solving.

- Develop a database of "best practices" for brownfields redevelopment (especially as coordinated with land-use planning), and make this widely available, perhaps via both Internet websites and print materials.

CHAPTER 8

Constraints to Incorporating Environmental Justice Principles in Land-Use Plans and Controls

Planners and planning officials may encounter particular legal and political constraints in incorporating environmental justice into land-use planning and regulation, as well as the usual limits brought about by a lack of either financial or staff resources.

JUDICIAL PROTECTIONS OF PRIVATE PROPERTY RIGHTS

The land-use regulatory model of environmental justice, while promising for many low-income communities of color, contains inherent limits. Among these limits are legal constraints on land-use regulation designed primarily to protect the private property rights of landowners. Uses of regulatory or eminent domain powers to pursue environmental justice planning policies conceivably could result in litigation and judicial review. Local governments mostly have broad discretion within certain bounds. While some experts perceive decreasing judicial deference to local land-use decision (Mandelker and Tarlock 1992; Wolf 1996; Dana 1997), recent U.S. Supreme Court decisions affirm local government use of eminent domain, planning processes, zoning, and other land-use regulatory tools, despite arguments that they infringed on private property rights (APA 2005; Bradley, Dowling, and Kendall 2006). Judicial opinions, legislation, and public opinion reflect a basic duality in which land-use planning and regulation exist: private property rights are valued and protected, yet necessarily subject to all kinds of limitations to protect neighboring property owners, the public, and the environment (Arnold 2002). While the field of permissible government action without risk of liability for compensation is quite broad, the precise point at which a given regulation might transgress judicially protected private property rights may depend on the underlying and surrounding facts of the case and on the court. Given the strong controls over private property that may be needed to ensure environmental justice in low-income communities of color, planning officials should be aware of the legal boundaries to their authority.

Given the strong controls over private property that may be needed to ensure environmental justice in low-income communities of color, planning officials should be aware of the legal boundaries to their authority.

There are four primary areas of constraints relevant to achievement of environmental justice in land-use plans, regulations, and decisions:

1. The reasonableness of the zoning decisions

2. The impact on the property owner's economically beneficial use of the property

3. A developer's expectations that zoning laws will not change once he or she has relied on initial approvals and begun the development; and

4. Rights to continue a previously permissible land-use once it has been prohibited.

First, the constitutional doctrine of *substantive due process* requires zoning to bear a real and substantial relationship to the public health, safety, morals, or welfare—the traditional police power justifications for regulation. The courts will strike down arbitrary, capricious, or unreasonable land-use controls or decisions (*Village of Euclid v. Ambler Realty Co.* 1926, 395). As discussed in Chapter 4 of this PAS Report, substantive due process claims often arise in situations of downzoning; the owner of the downzoned property will argue that the downzoning is arbitrary and capricious in its application to his or her property. The most important factors to courts in determining the validity of the downzoning are the reasons for the zoning change: Is it designed to stop a specific land-use proposal instead of resulting from preproposal comprehensive planning?; Are surrounding parcels are treated similarly?; and Is the degree to which the downzoning decreases the property's value and interferes with reasonable expectations about the use of the property (Arnold 1998, 125–26).

Second, the Takings Clause of the Fifth Amendment limits the government's regulation of land-use. The Supreme Court has developed several different tests depending on the government action respecting private property. The *Nollan* "essential nexus" and *Dolan* "rough proportionality" tests for the imposition of exactions are discussed in Chapter 4 of this report.

Physical occupation of private property (*Loretto v. Teleprompter Manhattan CATV Corporation* 1982) would rarely be relevant to the land-use regulation model of environmental justice and will not be discussed here. However, Supreme Court jurisprudence on regulatory takings is highly relevant. If a land-use regulation denies a property owner all of the economically viable use of his or her property, a taking has occurred and compensation is due, unless the property owner's rights never included the right to whatever activity is being regulated, such as a public nuisance (*Lucas v. South Carolina Coastal Council* 1992). If the landowner suffers a diminution in value less than 100 percent of the economically viable use of his or her property, courts will apply an ad hoc balancing test "that considers the economic effects of the regulation and the government's purpose" (*Penn Central Transportation Company v. City of New York*, 1978, 124).

Courts uphold zoning regulations that greatly restrict the use of private property far more than they declare such regulations to be takings (Wolf 1996, note 366). Many of the cases in which government agencies must compensate landowners involve total bans on development. However, some takings cases involve downzoning that both limits the use and diminishes the value of the property. When the property still has some significant value for the rezoned use, courts will find no taking, even if the diminution in value is substantial. Where the rezoned use is deemed economically unfeasible because the property is inappropriate for that use, though, a taking will have occurred. Often an important factor will be whether surrounding more-intensive uses (e.g., industrial or commercial uses or major roads or freeways) make a less-intensive zoning designation (e.g., single-family residential) unreasonable, therefore rendering the property relatively useless.

Therefore, as localities make zoning changes to achieve environmental justice, they should:

1) avoid using designations for private property that completely prevent development (e.g., open space designations);

2) seek compatible uses for contiguous parcels so that a single piece of land does not become a low-intensity island or peninsula in the midst of a sea of high-intensity uses;

3) explicitly connect any zoning changes to traditional state nuisance law to the extent possible; and

4) identify economically viable permissible uses for property subject to new zoning scheme.

In fact, the land-use model of environmental justice envisions local communities identifying productive, yet healthy, safe, and compatible uses for land, not merely prohibiting unwanted land uses.

Third, the doctrine of *vested rights and equitable estoppel* (estoppel being a legal bar to asserting or denying something when one has previously acted to the contrary) may prevent local governments from stopping a development proposal once the developer has obtained some approvals and relied on them in proceeding with the project (Rhodes and Sellers 1991; Hanes and Minchew 1989). The issue might arise, for example, when city officials learn of a proposed chemical recycling plant in the neighborhood and either rezone the property in question from light industrial use (which permits "recycling facilities") to commercial use or amend the zoning code text to prohibit chemical recycling plants in light industrial districts. If the developer has already received some city approvals (e.g., a site plan approval, a conditional use permit, or a building permit), at what point does he or she have a vested right in the zoning that existed at the time he or she obtained the initial approvals?

The land-use model of environmental justice envisions local communities identifying productive, yet healthy, safe, and compatible uses for land, not merely prohibiting unwanted land uses.

The area of vested rights and equitable estoppel has been termed "hopelessly muddled" (Hanes and Minchew 1989, 376–77, 382–83). The doctrine of vested rights, grounded in constitutional protections of private property rights against government interference, and the doctrine of equitable estoppel, or perhaps more precisely equitable zoning estoppel, grounded in equitable protections against unfair exercises of government zoning power, are distinct from each other only in theory; in practice, the concepts are treated interchangeably. In addition, the rules governing when a landowner has vested rights to proceed with development (or when a government regulator is estopped from preventing the development) vary considerably from state to state in ways that defy precise categorization. Conceptually, states can be divided into early vesting jurisdictions, which give the developer early certainty that zoning controls will not change in the midst of the multipermit approval process, and late vesting jurisdictions, which require the developer to have obtained one of the later permits given just before the final building phase takes place, such as a building permit (Hanes and Minchew 1989, 379–80). However, cases vary so much, not only from state to state, but even within states, that the conceptual distinctions do not closely match actual case outcomes in any predictable way.

A developer claiming vested rights or equitable zoning estoppel must establish three elements:

1) An official government act or omission that would suggest approval of the project

2) Good faith reliance on the government action

3) Substantial change in position or incurrence of extensive obligations and expenses toward developing the property (Hanes and Minchew 1989, 388–400; Rhodes and Seller 1991, 478–89).

Depending on the jurisdiction and the facts of the case, some of the following government approvals might result in vested rights:

- Approval of a site plan or planned unit development (PUD) when accompanied by a rezoning (e.g., to reflect the approved PUD use)

- Approval of a plat or subdivision site plan

- A conditional use (or special use) permit

- A preliminary permit (e.g., a rough grading, clearing, paving, foundation, or public improvement permit)

- Informal assurances and representations by local government officials

- Arguably, conditional zoning by which the developer commits to certain conditions in exchange for a specific zoning designation

If the developer, in good faith, relies on the requisite approvals by expending substantial amounts of money or making significant physical changes to the land, any subsequent zoning changes inconsistent with the earlier approvals will be invalid.

Therefore, planners and planning officials who seek zoning changes in low-income and minority neighborhoods might not be able to stop developments and land uses for which the developer has already received some initial approval(s). Planning staff and the city or county attorney's office will have to closely monitor which developers or property owners may have vested rights. Cities can avoid many of the problems with vested rights, however, if local officials or staff formally put a developer on notice of an intended or contemplated zoning change or other land-use controls to prevent the development, and if they give the notice before the developer has spent substantial sums on the project after having achieved some type of approval.

Fourth, as discussed in Chapter 4, the doctrine of *nonconforming uses* prevents a local government, when it makes a zoning change, from demanding the immediate discontinuance of a use that was lawful at the time of the zoning change, unless the use is a public nuisance. Existing uses allowed under the prior version of the zoning code, therefore, may be allowed to continue, but local officials can create a program of monitoring those uses and phasing-in their elimination over an amortization period.

STATE PREEMPTION OF LOCAL NIMBYISM

Another set of legal limits to land-use regulation as an environmental justice tool is state preemption of local land-use regulations and decisions that attempt to keep out Locally Unwanted Land Uses (LULUs). These laws are a response to the NIMBY ("Not In My Backyard") phenomenon, in which local residents mount powerful and effective campaigns to prevent LULUs from being located near them (Dear 1992; Heiman 1990; Delogu 1990). Environmental justice advocates have argued that NIMBYism by white and upper-income communities has contributed to the siting of noxious uses in less politically and economically powerful neighborhoods inhabited by low-income people and minorities (Gauna 1995, 32-33). However, just at the time when low-income and minority communities are trying to prevent LULUs and environmental hazards in their neighborhoods, state preemption laws designed to combat NIMBYism may hurt these environmental justice efforts.

Environmental justice advocates have argued that NIMBYism by white and upper-income communities has contributed to the siting of noxious uses in less politically and economically powerful neighborhoods inhabited by low-income people and minorities.

There are two basic types of LULUs subject to preemption in order to overcome local opposition to their siting (Mank 1995). The first is hazardous waste management facilities. Preemption of the siting of such facilities usually takes one of four courses:

1) "Super review," under which the private developer of a hazardous waste facility chooses a potential site and applies for a permit from a state agency. The agency reviews the environmental impacts and presents all applications meeting state environmental criteria to a special siting board that gathers public input, but this data gathering is often primarily designed to neutralize public opposition and fear.

2) "Site designation," under which the state agency—not the private developer—formulates a list of possible sites for hazardous waste facilities.

3) "Absolute prohibition," by which some states have expressly prohibited localities from using land-use requirements to burden the operation of hazardous waste facilities.

4) "Local control," which is followed primarily in California and Florida. Under this last approach, local regulation of hazardous waste facility siting is not preempted by state law, and localities are free to enact strict land-use regulations to keep out all hazardous waste sites.

The other type of LULU siting protected from local opposition by state statutory or judicial exemption is the siting of certain residential facilities, such as group homes for people with developmental disabilities, halfway houses, and low-income housing. State preemption laws of both types create some very real political and legal difficulties for local efforts to keep LULUs out of low-income and minority neighborhoods.

Even though zoning controls that prevent hazardous waste facilities or other LULUs in low-income or minority neighborhoods might be overridden by state laws, there are several important reasons to seek these zoning controls nonetheless. Zoning that does not permit a specified LULU (e.g., a hazardous waste facility) suggests to state regulators that the use is incompatible with surrounding land uses—a type of presumption in effect—whereas if the property is zoned to allow the LULU, state regulators are more likely to believe it is compatible with the neighboring land uses. If the local zoning allows the LULU, close scrutiny of its siting by any level of government agency may never occur, while an attempt by the locality to exclude it could put pressure on state regulators to find reasons to deny state permits. The zoning might also discourage potential developers or operators of LULUs from attempting to site the LULUs in that area. They might perceive that the neighborhood is politically active and opposed to such LULUs, which could lead to a costly and time-consuming approval process for them. They also might want to avoid a legal dispute to enforce the preemption and the negative publicity that can often accompany such battles.

Furthermore, the very process of developing land-use plans and regulations that reflect neighborhood goals and obtaining their enactment by local officials will tend to promote a more politically active and aware grassroots group. The group could mobilize more quickly and effectively to oppose a LULU proposal, even if decided at the state level, than if the community were forming a group for the first time upon learning of the specific proposal. In addition, there are many LULUs local residents might want to exclude and many beneficial land uses they might want to include, beyond the few land uses that are the subject of state control. In other words, many LULUs not subject to state preemption could be effectively precluded by local zoning. Even if a land-use plan will not effectively protect against every LULU, it should address some of the inequities in the distribution of land-use patterns, such as the high concentration of industrial and commercial uses in many low-income communities of color. Therefore, despite the obstacles presented by state preemption laws, local land-use regulation can be an effective environmental justice tool.

POLITICS AND PARTICIPATION

Although planners and planning officials are well aware of the political challenges and constraints of the land-use regulatory system, environmental justice issues pose some special political challenges.

Most essentially, disparities in power and participation are both contributing causes of environmental injustices and environmental justice problems in and of themselves, especially as defined by environmental justice groups and affected communities. Low-income people and people of color have historically had less power and participated less fully in land-use decision making than have others in local communities. Redistribution of relative power among various interests in local land-use policy can be conflict-ridden, contested, and downright ugly, often with local government officials caught in the middle.

The corollary of this observation is that owners, developers, and operators of industrial and commercial facilities and other LULUs have historically had more power than the residents of the communities in which their facilities were or became located. Some may cooperate with environmental justice planning and regulatory change, but others will fiercely oppose any such changes, especially if the changes are aimed to prevent or remove these facilities from mixed residential-industrial areas. The opponents of an environmental justice land-use policy initiative may lobby local elected and appointed officials, perhaps with considerable influence. They may seek to mobilize public opinion against the initiative by claiming it will cost jobs, economic growth, and tax revenues. They may launch appeals to concepts of fairness, claiming they were there first and area residents voluntarily

Disparities in power and participation are both contributing causes of environmental injustices and environmental justice problems in and of themselves, especially as defined by environmental justice groups and affected communities.

assumed the risks of living in what could be considered essentially industrial areas. They may lobby state legislators for legislation to preempt local environmental justice efforts. They may litigate, challenging the validity of regulatory changes or permit denials or claiming compensation from government regulators. In addition, owners of industrially or commercially zoned property will often oppose downzoning of those parcels, the imposition of additional controls via overlay districts or performance zoning, and demands of exactions. These landowners are likely to have financial and political capital to spend in seeking to defeat an environmental justice land-use plan.

Residents of low-income and minority neighborhoods might not cooperate with local planners' efforts to incorporate environmental justice considerations into land-use plans and actions. First, they may be angry and distrustful after a long history of power disparities and environmental and land-use injustices. They may not have much faith in land-use planning and regulation to achieve a clean, healthy, vibrant, and just environment for them, and may distrust local government generally. They also may not be in any mood to cooperate with decision makers or commercial and industrial occupants of their neighborhoods, viewing any degree of cooperation or collaborative problem solving as essentially capitulation. Protest and opposition may be their methods of seeking environmental justice. Second, they may fail to participate in any significant numbers. The reasons can range from structural barriers to participation (e.g., meeting times and locations, language used, lack of information and expertise, structure of meetings, feelings of intimidation or distrust), to personal constraints (e.g., work and family obligations, disabilities and health problems, educational and informational gaps), to apathy, feelings of inefficacy, or plain unawareness of the issues in their community. In addition, even if there is widespread initial participation, maintaining interest and involvement in long-term planning processes, regulatory changes, permit decisions, and implementation and enforcement is an enormous challenge, at best. Third, neighborhood or area residents may not be united in their goals and may disagree about general policies and specific land-use compatibility issues. Some residents might embrace one or more LULUs, while others strongly oppose them. Even more likely, different community-based groups may compete with one another to speak for the community, to influence policy outcomes, and to obtain limited resources. Sadly, within a neighborhood or community experiencing environmental injustice with a number of significant issues to address, one grassroots group may refuse to participate if a competing group participates.

Planners might also encounter resistance or opposition from elected or appointed officials or government administrators. These government leaders might perceive an environmental justice land-use policy initiative as politically risky, too controversial, or perhaps simply not politically beneficial enough to warrant its pursuit. They might favor other priorities for the expenditure of scarce resources, including public funds, staff time, and places on the policy agenda. They might regard changes to existing industrial or commercial zoning as politically or fiscally inconvenient, especially when these uses cannot be relocated to higher-income, lower-minority areas without political conflict. Indeed, many local governments engage in "fiscal zoning," favoring industrial and commercial uses because these uses generate tax revenues without creating expensive demands for local services in the way that single-family residences do, particularly due to public school costs (Ellickson and Tarlock 1981, 738–40). Single-family residential neighborhoods, particularly if occupied by upper-income people, however, are desirable for other reasons than a pure analysis of marginal costs and revenues would indicate, but cities and counties might offset the

Government leaders might perceive an environmental justice land-use policy initiative as politically risky, too controversial, or perhaps simply not politically beneficial enough to warrant its pursuit.

Increasingly, the idea that poor people and people of color should bear a disproportionately high incidence of health risks, environmental harms, and distressed communities simply because of political or economic expediency is morally repugnant to many.

costs of these neighborhoods by reducing expenditures on older neighborhoods where industrial and commercial uses have intruded; and, generally, these are low-income and minority neighborhoods. Therefore, fiscal zoning practices can have a double negative effect on low-income communities of color: 1) the attraction of industrial and commercial uses to those areas, and 2) pressures on local governments to decrease public spending on physical infrastructure, schools, and other public services in those areas.

The manner in which issues, information, and ideas are presented may have a significant influence on the degree of support by local officials and the public. Planners should consider:

- using easy-to-understand visual representations of harms and environmental/land-use conditions in the community;

- articulating as clearly as possible the purposes an initiative will serve;

- identifying the opportunities presented by improving environmental and land-use conditions for the least advantaged or powerful communities (e.g., improved economy, improved public health and safety, improved civic commitment and citizen satisfaction); and

- presenting concrete projects and tools to accomplish the initiative's goals.

Finally, the nature of the land-use planning and regulatory model of environmental justice requires continual involvement in, and monitoring of, implementation. Developers, landowners, and LULU operators may seek conditional use permits, variances, and rezonings, among other changes or exceptions to whichever policies and regulations have incorporated environmental justice principles. An all-too-common policy failure is the adoption of a broad reform initiative subsequently undermined by exceptions and variances, nonimplementation, poor enforcement, and lack of resources, resulting in a merely symbolic or token policy.

Despite all of these potential political limitations, they are not insurmountable. This discussion paints a worst-case scenario; in many circumstances, support for environmental justice reforms to land-use policies and regulations is relatively balanced with, or even greater than, opposition. Awareness of environmental injustices and land-use inequities is growing. Increasingly, the idea that poor people and people of color should bear a disproportionately high incidence of health risks, environmental harms, and distressed communities simply because of political or economic expediency is morally repugnant to many. Finally, the case studies in this PAS Report highlight examples of the many localities incorporating environmental justice principles and practices into land-use. In short, environmentally equitable and sustainable land-use practices are growing, despite the political constraints.

Afterword

The opportunities to incorporate environmental justice principles into land-use planning and regulation are abundant. This is so because land-use planning and regulation offer the promise of a clean, healthy, vibrant, and just environment for all people. Land-use planners and planning officials have many tools and techniques at their disposal to make this promise a reality in their communities. The scope of issues affected by environmental injustices is broad, and the theory and practice of planning for environmental justice and equitable land-use practices will continue to develop. After all, so much that is important in our society— from the health of children to the strength of our social fabric—is at stake.

Executive Order 12898:
Federal Actions to Address Environmental Justice in Minority Populations and Low-Income Populations

President of the United States
Executive Order 12898
February 11, 1994
Federal Actions to Address Environmental Justice in
Minority Populations and Low-Income Populations

FEDERAL REGISTER, VOL. 59, No. 32
Presidential Documents
59 FR 7629

DATE: Wednesday, February 16, 1994

By the authority vested in me as President by the Constitution and the laws of the United States of America, it is hereby ordered as follows:

Section 1-1. Implementation.

1-101. Agency Responsibilities.

To the greatest extent practicable and permitted by law, and consistent with the principles set forth in the report on the National Performance Review, each Federal agency shall make achieving environmental justice part of its mission by identifying and addressing, as appropriate, disproportionately high and adverse human health or environmental effects of its programs, policies, and activities on minority populations and low-income populations in the United States and its territories and possessions, the District of Columbia, the Commonwealth of Puerto Rico, and the Commonwealth of the Mariana Islands.

1-102. Creation of an Interagency Working Group on Environmental Justice.

a) Within 3 months of the date of this order, the Administrator of the Environmental Protection Agency ("Administrator") or the Administrator's designee shall convene an interagency Federal Working Group on Environmental Justice ("Working Group"). The Working Group shall comprise the heads of the following executive agencies and offices, or their designees: (a) Department of Defense; (b) Department of Health and Human Services; (c) Department of Housing and Urban Development; (d) Department of Labor; (e) Department of Agriculture; (f) Department of Transportation; (g) Department of Justice; (h) Department of the Interior; (i) Department of Commerce; (j) Department of Energy; (k) Environmental Protection Agency; (l) Office of Management and Budget; (m) Office of Science and Technology Policy; (n) Office of the Deputy Assistant to the President for Environmental Policy; (o) Office of the Assistant to the President for Domestic Policy; (p) National Economic Council; (q) Council of Economic Advisers; and (r) such other Government officials as the President may designate. The Working Group shall report to the President through the Deputy Assistant to the President for Environmental Policy and the Assistant to the President for Domestic Policy.

(b) The Working Group shall:

1. provide guidance to Federal agencies on criteria for identifying disproportionately high and adverse human health or environmental effects on minority populations and low-income populations;

2. coordinate with, provide guidance to, and serve as a clearinghouse for, each Federal agency as it develops an environmental justice strategy as required by section 1-103 of this order, in order to ensure that the administration, interpretation and enforcement of programs, activities and policies are undertaken in a consistent manner;

3. assist in coordinating research by, and stimulating cooperation among, the Environmental Protection Agency, the Department of Health and Human Services, the Department of Housing and Urban Development, and other agencies conducting research or other activities in accordance with section 3-3 of this order;

4. assist in coordinating data collection, required by this order; 59 FR 7630

5. examine existing data and studies on environmental justice;

6. hold public meetings as required in section 5-502(d) of this order; and

7. develop interagency model projects on environmental justice that evidence cooperation among Federal agencies.

1-103. Development of Agency Strategies.

(a) Except as provided in section 6-605 of this order, each Federal agency shall develop an agency-wide environmental justice strategy, as set forth in subsections (b)-(e) of this section that identifies and addresses disproportionately high and adverse human health or environmental effects of its programs, policies, and activities on minority populations and low-income populations. The environmental justice strategy shall list programs, policies, planning and public participation processes, enforcement, and/or rulemakings related to human health or the environment that should be revised to, at a minimum:

    (1) promote enforcement of all health and environmental statutes in areas with minority populations and low-income populations;

    (2) ensure greater public participation;

    (3) improve research and data collection relating to the health of and environment of minority populations and low-income populations; and

    (4) identify differential patterns of consumption of natural resources among minority populations and low-income populations. In addition, the environmental justice strategy shall include, where appropriate, a timetable for undertaking identified revisions and consideration of economic and social implications of the revisions.

(b) Within 4 months of the date of this order, each Federal agency shall identify an internal administrative process for developing its environmental justice strategy, and shall inform the Working Group of the process.

(c) Within 6 months of the date of this order, each Federal agency shall provide the Working Group with an outline of its proposed environmental justice strategy.

(d) Within 10 months of the date of this order, each Federal agency shall provide the Working Group with its proposed environmental justice strategy.

(e) Within 12 months of the date of this order, each Federal agency shall finalize its environmental justice strategy and provide a copy and written description of its strategy to the Working Group. During the 12 month period from the date of this order, each Federal agency, as part of its environmental justice strategy, shall identify several specific projects that can be promptly undertaken to address particular concerns identified during the development of the proposed environmental justice strategy, and a schedule for implementing those projects.

(f) Within 24 months of the date of this order, each Federal agency shall report to the Working Group on its progress in implementing its agency-wide environmental justice strategy.

(g) Federal agencies shall provide additional periodic reports to the Working Group as requested by the Working Group.

1-104. Reports to the President.

Within 14 months of the date of this order, the Working Group shall submit to the President, through the Office of the Deputy Assistant to the President for Environmental Policy and the Office of the Assistant to the President for Domestic Policy, a report that describes the implementation of this order, and includes the final environmental justice strategies described in section 1-103(e) of this order.

Section 2-2. Federal Agency Responsibilities for Federal Programs.

Each Federal agency shall conduct its programs, policies, and activities that substantially affect human health or the environment, in a manner that ensures that such programs, policies, and activities do not have the effect of excluding persons (including populations) from participation in, denying persons (including populations) the benefits of, or subjecting persons (including populations) to discrimination under, such programs, policies, and activities, because of their race, color, or national origin. 59 FR 7631.

Section 3-3. Research, Data Collection, and Analysis.

3-301. Human Health and Environmental Research and Analysis.

(a) Environmental human health research, whenever practicable and appropriate, shall include diverse segments of the population in epidemiological and clinical studies, including segments at high risk from environmental hazards, such as minority populations, low-income populations and workers who may be exposed to substantial environmental hazards.

(b) Environmental human health analyses, whenever practicable and appropriate, shall identify multiple and cumulative exposures.

(c) Federal agencies shall provide minority populations and low-income populations the opportunity to comment on the development and design of research strategies undertaken pursuant to this order.

3-302. Human Health and Environmental Data Collection and Analysis.

To the extent permitted by existing law, including the Privacy Act, as amended (5 U.S.C. section 552a):

(a) each Federal agency, whenever practicable and appropriate, shall collect, maintain, and analyze information assessing and comparing environmental and human health risks borne by populations identified by race, national origin, or income. To the extent practical and appropriate, Federal agencies shall use this information to determine whether their programs, policies, and activities have disproportionately high and adverse human health or environmental effects on minority populations and low-income populations;

(b) In connection with the development and implementation of agency strategies in section 1-103 of this order, each Federal agency, whenever practicable and appropriate, shall collect, maintain and analyze information on the race, national origin, income level, and other readily accessible and appropriate information for areas surrounding facilities or sites expected to have a substantial environmental, human health, or economic effect on the surrounding populations, when such facilities or sites become the subject of a substantial Federal environmental administrative or judicial action. Such information shall be made available to the public, unless prohibited by law; and

(c) Each Federal agency, whenever practicable and appropriate, shall collect, maintain, and analyze information on the race, national origin, income level, and other readily accessible and appropriate information for areas surrounding Federal facilities that are:

(1) subject to the reporting requirements under the Emergency Planning and Community Right-to-Know Act, 42 U.S.C. section 11001-11050 as mandated in Executive Order No. 12856; and

(2) expected to have a substantial environmental, human health, or economic effect on surrounding populations. Such information shall be made available to the public, unless prohibited by law.

(d) In carrying out the responsibilities in this section, each Federal agency, whenever practicable and appropriate, shall share information and eliminate unnecessary duplication of efforts through the use of existing data systems and cooperative agreements among Federal agencies and with State, local, and tribal governments.

Section 4-4. Subsistence Consumption of Fish and Wildlife

4-401. Consumption Patterns

In order to assist in identifying the need for ensuring protection of populations with differential patterns of subsistence consumption of fish and wildlife, Federal agencies, whenever practicable and appropriate, shall collect, maintain, and analyze information on the consumption patterns of populations who principally rely on fish and/or wildlife for subsistence. Federal agencies shall communicate to the public the risks of those consumption patterns. 59 FR7632

4-402. Guidance

Federal agencies, whenever practicable and appropriate, shall work in a coordinated manner to publish guidance reflecting the latest scientific information available concerning methods for evaluating the human health risks associated with the consumption of pollutant-bearing fish or wildlife. Agencies shall consider such guidance in developing their policies and rules.

Section 5-5. Public Participation and Access to Information.

(a) The public may submit recommendations to Federal agencies relating to the incorporation of environmental justice principles into Federal agency programs or policies. Each Federal agency shall convey such recommendations to the Working Group.

(b) Each Federal agency may, whenever practicable and appropriate, translate crucial public documents, notices, and hearings relating to human health or the environment for limited English speaking populations.

(c) Each Federal agency shall work to ensure that public documents, notices, and hearings relating to human health or the environment are concise, understandable, and readily accessible to the public.

(d) The Working Group shall hold public meetings, as appropriate, for the purpose of fact-finding, receiving public comments, and conducting inquiries concerning environmental justice. The Working Group shall prepare for public review a summary of the comments and recommendations discussed at the public meetings.

Section 6-6. General Provisions.

6-601. Responsibility for Agency Implementation.

The head of each Federal agency shall be responsible for ensuring compliance with this order. Each Federal agency shall conduct internal reviews and take such other steps as may be necessary to monitor compliance with this order.

6-602. Executive Order No. 12250.

This Executive order is intended to supplement but not supersede Executive Order No. 12250, which requires consistent and effective implementation of various laws prohibiting discriminatory practices in programs receiving Federal financial assistance. Nothing herein shall limit the effect or mandate of Executive Order No. 12250.

6-603. Executive Order No. 12875.

This Executive order is not intended to limit the effect or mandate of Executive Order No. 12875.

6-604. Scope.

For purposes of this order, Federal agency means any agency on the Working Group, and such other agencies as may be designated by the President, that conducts any Federal program or activity that substantially affects human health or the environment. Independent agencies are requested to comply with the provisions of this order.

6-605. Petitions for Exemptions.

The head of a Federal agency may petition the President for an exemption from the requirements of this order on the grounds that all or some of the petitioning agency's programs or activities should not be subject to the requirements of this order.

6-606. Native American Programs.

Each Federal agency responsibility set forth under this order shall apply equally to Native American programs. In addition, the Department of the Interior, in coordination with the Working Group, and, after consultation with tribal leaders, shall coordinate steps to be taken pursuant to this order that address Federally-recognized Indian Tribes.

6-607. Costs.

Unless otherwise provided by law, Federal agencies shall assume the financial costs of complying with this order.

6-608. General.

Federal agencies shall implement this order consistent with, and to the extent permitted by, existing law.

6-609. Judicial Review.

This order is intended only to improve the internal management of the executive branch and is not intended to, nor does it create any right, benefit, or trust responsibility, substantive or procedural, [*7633] enforceable at law or equity by a party against the United States, its agencies, its officers, or any person. This order shall not be construed to create any right to judicial review involving the compliance or noncompliance of the United States, its agencies, its officers, or any other person with this order.

William J. Clinton
The White House
February 11, 1994

Summaries of the American Planning Association's Policy Guide Provisions Relating to Environmental Justice

Policy Guide on Smart Growth 2002

[Editor's Note: The full text of this policy guide is available at www.planning.org/policyguides/smartgrowth.htm]

This policy guide urges planning for smart growth to achieve communities that:

- have a unique sense of community and place;

- preserve and enhance valuable natural and cultural resources;

- equitably distribute the costs and benefits of development;

- expand the range of transportation, employment, and housing choices in a fiscally responsible manner;

- value long-range, regional considerations of sustainability over short term incremental geographically isolated actions; and

- promote public health and healthy communities.

Many of the general principles of smart growth and specific policy motions with a regional focus are relevant to low-income communities of color. Specific policies include those in the categories of: 1) planning structure, process, and regulation; 2) transportation and land use; 3) regional management and fiscal efficiency; and 4) environmental protection and land conservation. For example, environmental justice principles are served by linkages between land use and transportation choices, planning for alternatives to automotive transportation, investment in land reuse and existing urban infrastructure, and green design practices if all of these, and other smart growth strategies, occur in low-income and minority communities.

However, a fifth category of specific policies addresses social equity and community building, which are directly related to environmental justice concerns. These policies include:

- a sustained and focused initiative in federal, state, and local public policy to reverse the general decline of urban neighborhoods and the trend toward isolated, concentrated poverty through strategies that promote reinvestment within urban communities;

- increased social, economic, and racial equity in our communities;

- increased federal funding of community development to remedy inequities;

- inclusion of input from all segments of the population in the planning process;

- planning and development decisions that do not unfairly burden economically disadvantaged groups;

- federal and state policies and programs that encourage mixed income neighborhoods as the foundation for healthy regions, including requirements for the provision of affordable housing in all new-growth areas or through reinvestment in core communities;

- the enhancement of public educations systems which are an essential component of community building in urban, suburban, and rural areas, and which ensure that children have an opportunity for an excellent education in existing communities;

- strategies that increase neighborhoods that are economically and socially diverse;

- planning that identifies the transportation, housing, employment, education, and other needs of population change, both with respect to the total number of people expected to reside in a region but also with respect to population groups with special needs such as the elderly, school children, or people of diverse cultures.

Policy Guide on Planning for Sustainability 2000

[Editor's Note: The full text of this policy guide is available at www.planning.org/policyguides/ sustainability.htm]

This policy guide encourages policies of sustainability, including:

- sustaining communities as good places to live and as places that offer economic and other opportunities to their inhabitants;
- sustaining the values of society, such as individual liberty and democracy;
- sustaining the biodiversity of the natural environment, both for the contribution that it makes to the quality of human life and for its own inherent value;
- sustaining the ability of natural systems to provide the life-supporting "services" that are rarely counted by economists, but which have recently been estimated to be worth nearly as much as the total gross human economic product.

Many of the policies relate to principles of environmental justice, because low-income and minority communities may be affected, often disproportionately, by unsustainable practices. Moreover, generally applicable goals like reduced use of chemicals and synthetic compounds, minimization or elimination of extraction of underground substances, avoidance of continued sprawl through compact and mixed-use development, and improved protections of water quality will benefit low-income and minority communities that are burdened by the impacts of current practices.

Nonetheless, the policy guide also directly addresses issues of social equity. It identifies racial and economic segregation, poverty, and inequality of opportunity as problems of unsustainability. It also encourages:

- planning policies and legislation at all levels of government that seek to equitably protect public health, safety and welfare, and to incorporate the needs of those currently disenfranchised in the process;

- planning policies and legislation encouraging participatory and partnership approaches to planning, including planning for sustainability, integrally involving local community residents in setting the vision for and developing plans and actions for their communities and regions, with special emphases on establishing avenues for meaningful participation in decision making by historically disadvantaged people and on basing decisions on community visions and plans;

- provision of affordable, efficient transportation alternatives for everyone, especially low-income households, elders, and others comprising 30% of the national population that cannot or do not own cars;

- provision of communities that are socially cohesive, reduce isolation, foster community spirit, share resources, and have divers occupants in terms of age and social and cultural groups;

- provision of housing that is affordable to a variety of income groups within the same community and that is located near employment centers:

- policies facilitating and encouraging businesses that meet human needs fairly and efficiently by fulfilling local and employment and consumer needs without degrading the environment, promoting financial and social equity in the workplace, creating vibrant community-based economies, providing employment opportunities that allow people economic self-determination and environmental health, and supporting locally-based agriculture;

- wastewater practices that clean, conserve, and reuse wastewater at the site, neighborhood, or community level, thus reducing the need for regional processing facilities; and

- fair and equitable growth management policies maintaining diversity in local populations and economies.

Policy Guide on Neighborhood Collaborative Planning 1998

[Editor's Note: The full text of this policy guide is available at www.planning.org/policyguides/ neighborhood.htm]

This policy guide encourages comprehensive planning that integrates community-wide planning with neighborhood collaborative planning. The guide states: "Neighborhoods should be recognized as building blocks of overall community development. Local officials and planners must heed opinions and suggestions of people and groups within the neighborhood to create a framework that will enable plans to have a greater chance of being supported and implemented, not only at the neighborhood level, but at the municipal, regional and even state levels."

The guide contains numerous findings and policy suggestions to improve neighborhood planning, including in low-income and minority neighborhoods. Three specific policy positions address the particular needs in low-income neighborhoods of color:

- Neighborhood-based coalitions that assist in the development of individual neighborhood organizations, articulate neighborhood views on community wide issues, and facilitate coordination in the planning process should be encouraged and supported by local government.

- Advocacy planning for neighborhoods should be accepted as a legitimate role for professional planners, both publicly and privately employed.

- To be effective in many cases, neighborhood planning needs to be beyond addressing the physical conditions of the area and also examine issues of social equity.

List of References

Books, articles, websites, and similar secondary sources

AASHTO (American Association of State and Highway Transportation Officials), Center for Environmental Excellence. 2006. "Environmental Justice." http://environment. transportation.org/environmental_issues/environmental_justice.

Acker, Frederick W. 1991. "Performance Zoning." *Notre Dame Law Review* 67: 363.

Agyeman, Julian. 2005. *Sustainable Communities and the Challenge of Environmental Justice.* New York: NYU Press.

Alves, Antonio, Trish Settles, and Jason Webb. 1995. "Environmentalism in the Dudley Street Neighborhood." *Virginia Environmental Law Journal* 14: 735.

Anderson, Andy B., et al. 1994. "Environmental Equity: Evaluating TSDF Siting Over the Past Two Decades." *Waste Age* (July) 84.

Anderson, Jerry L., and Erin Sass. 2004. "Is the Wheel Unbalanced: A Study of Bias on Zoning Boards." *Urban Lawyer* 36: 447.

Anderson, Jerry L., and Daniel Luebbering. 2006. "Zoning Bias II: A Study of Oregon's Zoning Commission Composition Restrictions." *Urban Lawyer* 38: 63.

APA (American Planning Association). 1979. "Policies and Commentary." *Planning*, July, 24B.

-----. 1994. *Planning and Community Equity: A Component of APA's Agenda for America's Communities Program.* Chicago: American Planning Association.

-----. 1998. *Policy Guide on Neighborhood Collaborative Planning.* Chicago: American Planning Association.

-----. 2000. *Policy Guide on Planning for Sustainability.* Chicago: American Planning Association.

-----. 2002. *Policy Guide on Smart Growth.* Chicago: American Planning Association.

-----. 2004. *Policy Guide on Public Redevelopment.* Chicago: American Planning Association.

-----. 2005. *Four Supreme Court Land-Use Decisions of 2005.* Planning Advisory Service Report No. 535. Chicago.

Arnold, Craig Anthony (Tony). 1998. "Planning Milagros: Environmental Justice and Land-use Regulation." *Denver University Law Review* 76: 1.

-----. 2000. "Land-use Regulation and Environmental Justice." *Environmental Law Reporter* 30: 10395.

-----. 2002. "Land-use Justice." *Projections: The Massachusetts Institute of Technology Journal of Planning* 3, no. 2: 32.

-----. 2005. "Is Wet Growth Smarter Than Smart Growth?: The Fragmentation and Integration of Land-use and Water." *Environmental Law Reporter* 35: 10152.

-----. 2006. "For the Sake of Watersheds: Land Conservation and Watershed Protection." *Sustain: A Journal of Environmental and Sustainability Issues* 14 (Spring/Summer): 16.

Arnstein, Sherry R. 1969. "A Ladder of Citizen Participation." *Journal of the American Institute of Planners* 35, no. 4: 216.

Baden, Brett, and Don Coursey. 1997. *The Locality of Waste Sites within the City of Chicago: A Demographic, Social, and Economic Analysis.* Chicago: The Irving B. Harris Graduate School of Public Policy Studies, University of Chicago Working Paper Series 97-2.

Barnett, Jonathan. 2003. *Redesigning Cities: Principles, Practice, Implementation.* Chicago: Planners Press.

Bates, Timothy. 2006. "The Urban Development Potential of Black-Owned Businesses." *Journal of the American Planning Association* 72(2): 227.

Beatley, Timothy. 1994. *Ethical Land-use: Principles of Policy and Planning*. Baltimore: The Johns Hopkins University Press.

Been, Vicki. 1991. "'Exit' as a Constraint on Land-use Exactions: Rethinking the Unconstitutional Conditions Doctrine." *Columbia Law Review* 91: 473.

-----. 1993. "What's Fairness Got to Do With It? Environmental Justice and the Siting of Locally Undesirable Land uses." *Cornell Law Review* 78: 1001.

-----. 1994. "Locally Undesirable Land uses in Minority Neighborhoods: Disproportionate Siting or Market Dynamics?" *Yale Law Journal* 103: 1383.

-----. 1995. "Analyzing Evidence of Environmental Justice." *Journal of Land-use and Environmental Law* 11: 1.

Been, Vicki, and Francis Gupta. 1997. "Coming to the Nuisance or Going to the Barrios? A Longitudinal Analysis of Environmental Justice Claims." *Ecology Law Quarterly* 24: 1.

Berube, Alan, and Bruce Katz. 2005. *Katrina's Window: Confronting Concentrated Poverty Across America*. Washington, D.C.: The Brookings Institution.

Binder, Denis. 1995. "Index of Environmental Justice Cases." *Urban Lawyer* 27: 163.

-----. 2000. "Environmental Justice Index II." *Chapman Law Review* 3: 309.

-----. 2005. "Environmental Justice Index III." *Environmental Law Reporter* 35: 10605.

Binder, Denis, et al. 2001. "A Survey of Federal Agency Responses to President Clinton's Executive Order 12898 on Environmental Justice." *Environmental Law Reporter* 21: 11133.

Blackwell, Angela Glover. 2001. "Promoting Equitable Development." *Indiana Law Review* 34: 1,273.

Blackwell, Robert W. 1989. "Overlay Zoning, Performance Standards, and Environmental Protection After *Nollan*." *Boston College Environmental Affairs Law Review* 16: 615.

Bobrowski, Mark. 1995. "Scenic Landscape Protection Under the Police Power." *Boston College Environmental Affairs Law Review* 22: 697.

Bond, Kenneth W. 1976. "Toward Equal Delivery of Municipal Services in the Central Cities." *Fordham Urban Law Journal* 4: 263.

Bonham, Jr., J. Blaine, Gerri Spilka, and Darl Rastorfer. 2002. *Old Cities/Green Cities: Communities Transform Unmanaged Land*. APA PAS Report No. 506/507. Chicago: American Planning Association.

Bonorris, Steven, et al. 2004. *Environmental Justice for All: A Fifty-State Survey of Legislation, Policies, and Initiatives*. American Bar Association and the University of California Hastings College of Law.

Boston, Thomas D. 2005. "The Effects of Revitalization on Public Housing Residents." *Journal of the American Planning Association* 71(4): 393.

Bourassa, Steven C. 2006. "The Community Land Trust as a Highway Environmental Impact Mitigation Tool." *Journal of Urban Affairs* 28 (4): 399.

Bradley, Jennifer, Timothy Dowling, and Douglas Kendall. 2006. *The Good News About Takings*. Chicago: APA Planners Press.

Bradman, Asa, et al. 2005. "Association of Housing Disrepair Indicators with Cockroach and Rodent Infestation in a Cohort of Pregnant Latina Women and Their Children." *Environmental Health Perspectives* 113(12): 1,795.

Brook, Michael P. 2002. *Planning Theory for Practitioners*. Chicago: American Planning Association.

Brown, Ken. 2006. "Point: Race, Poverty and Redevelopment." *Brownfield News*. www.brownfieldnews.com. June 6.

Bryant, Bunyan. ed. 1995. *Environmental Justice: Issues, Policies, and Solutions*. Washington, DC: Island Press.

Bryant, Bunyan, and Paul Mohai. eds. 1992. *Race and the Incidence of Environmental Hazards: A Time for Discourse*. Boulder, Colo.: Westview Press.

Bullard, Robert D. 1987. *Invisible Houston: The Black Experience in Boom and Bust*. College Station: Texas A & M University Press.

-----. 1990. *Dumping in Dixie: Race, Class, and Environmental Quality*. Boulder, Colo.: Westview Press.

-----. ed. 1993. *Confronting Environmental Racism: Voices from the Grassroots*. Boston: South End Press.

-----. ed. 1994. *Unequal Protection: Environmental Justice and Communities of Color*. San Francisco: Sierra Club Books.

-----. ed. 2005. *The Quest for Environmental Justice: Human Rights and the Politics of Pollution*. San Francisco: Sierra Club Books.

Bullard, Robert D., and Glenn S. Johnson. eds. 1997. *Just Transportation: Dismantling Race and Class Barriers to Mobility*. Stony Creek, Conn.: New Society Publishers.

Bullard, Robert D., Glenn S. Johnson, and Angel O. Torres. eds. 2000. *Sprawl City: Race, Politics, and Planning in Atlanta*. Washington, D.C.: Island Press.

-----. 2002. "Transportation Justice for All: Addressing Equity in the 21st Century." *Second National People of Color Environmental Leadership Summit – Summit II Resource Paper Series*. Washington, D.C.: Summit II National Office.

Burke, Edmund M. 1979. *A Participatory Approach to Urban Planning*. New York: Human Sciences Press.

California Environmental Protection Agency and California Air Resources Board. 2005. *Air Quality and Land Use Handbook: A Community Health Perspective*. Sacramento.

Callies, David L., et al. 1994. *Cases and Materials on Land-use*. 2d edition. St. Paul, Minn.: West Publishing.

Camacho, Alejandro Esteban. 2005. "Mustering the Missing Voices: A Collaborative Model for Fostering Equality, Community Involvement and Adaptive Planning in Land Use Decisions, Installment Two." *Stanford Environmental Law Journal* 24: 269.

Campbell, Scott. 1996. "Green Cities, Growing Cities, Just Cities? Urban Planning and the Contradictions of Sustainable Development." *Journal of the American Planning Association* 62, no. 3: 296.

Cashin, Sharyll L. 2004. *The Failures of Integration: How Race and Class Are Undermining the American Dream*. New York: Public Affairs Books.

Catanese, Anthony James. 1978. "Learning by Comparison: Lessons from Experience." In *Personality, Politics, and Planning*, edited by Anthony James Catanese and W. Paul Farmer. Chicago: American Planning Association.

-----. 1984. *The Politics of Planning and Development*. Beverly Hills, Calif.: Sage.

Center for Progressive Regulation. 2005. *An Unnatural Disaster: The Aftermath of Hurricane Katrina*. Washington, D.C.: Center for Progressive Regulation.

Centner, Terence J., et al. 1996. "Environmental Justice and Toxic Releases: Establishing Evidence of Discriminatory Effect Based on Race and Not Income." *Wisconsin Environmental Law Journal* 3: 119.

Chapin, Jr., F. Stuart, and Edward J. Kaiser. 1979. *Urban Land-use Planning*. 3d edition. Champaign, Ill.: University of Illinois Press.

The City Project. 2007a. "Urban Parks." www.cityprojectca.org.

-----. 2007b. "Los Angeles River." www.cityprojectca.org.

Cole, Luke W. 1992. "Empowerment as the Key to Environmental Protection: The Need for Environmental Poverty Law." *Ecology Law Quarterly* 19: 619.

-----. 1994. "Environmental Justice Litigation: Another Stone in David's Sling." *Fordham Urban Law Journal* 21: 523.

Cole, Luke W., and Sheila R. Foster. 2001. *From the Ground Up: Environmental Racism and the Rise of the Environmental Justice Movement*. New York: New York University Press.

Collin, Robert W. 1992. "Environmental Equity: A Law and Planning Approach to Environmental Racism." *Virginia Environmental Law Journal* 11: 495.

-----. 1994. "Review of the Legal Literature on Environmental Racism, Environmental Equity, and Environmental Justice." *Journal of Environmental Law and Litigation 9*: 121.

Collin, Robin Morris, and Robert Collin. 2001. "Sustainability and Environmental Justice: Is the Future Clean and Black?" *Environmental Law Reporter* 31: 10968.

Collin, Robin Morris, and Robert Collin. 2005. "Environmental Reparations." In *The Quest for Environmental Justice: Human Rights and the Politics of Pollution*, edited by Robert D. Bullard. San Francisco: Sierra Club Books.

Comfort, Louis K. 2006. "Cities at Risk: Hurricane Katrina and the Drowning of New Orleans." *Urban Affairs Review* 41 (4): 501–16.

Colorado People's Environmental and Economic Network (COPEEN). 2000. "Zoned Out." www.copeen.org/ej/zoned_out.htm, visited March 21, 2005.

Congress for New Urbanism. 2004. *Codifying New Urbanism: How to Reform Municipal Land Development Regulations*. Planning Advisory Service Report No. 526. Chicago: American Planning Association.

Conzo, Jessica. 2006. "Mapping New York City's Cultural Assets: A Pilot Project of East Harlem." Master's Thesis. New York: The New School University.

Costanza, Robert, et al. 1997. "The Value of the World's Ecosystem Services and Natural Capital." *Nature* 387: 253.

Crowley, Sheila. 2003. "The Affordable Housing Crisis: Residential Mobility of Poor Families and School Mobility of Poor Children." *Journal of Negro Education* 72(1): 22.

Cutter, Susan L., et al. 2003. "Social Vulnerability to Environmental Hazards." *Social Science Quarterly* 84 (1): 242–61.

Daily, Gretchen C. ed. 1997. *Nature's Services: Societal Dependence on Natural Ecosystems*. Washington, D.C.: Island Press.

Dana, David A. 1997. "Land-use Regulation in an Age of Heightened Scrutiny." *North Carolina Law Review* 75: 1243.

Daniels, Tom, and Katherine Daniels. 2003. *The Environmental Planning Handbook for Sustainable Communities and Regions*. Chicago: American Planning Association.

Davidoff, Paul. 1965. "Advocacy and Pluralism in Planning." *Journal of the American Institute of Planners* 31, no. 4: 48.

Davis, Todd S. 2002. *Brownfields: A Comprehensive Guide to Redeveloping Contaminated Property*. Chicago: American Bar Association.

Day, Kristen. 2006. "Active Living and Social Justice: Planning for Physical Activity in Low-income, Black, and Latino Communities." *Journal of the American Planning Association* 72, no. 1: 88.

de Souza Briggs, Xavier. ed. 2005. *The Geography of Opportunity: Race and Housing Choice in Metropolitan America*. Washington, DC: Brookings Institution Press.

Dear, Michael. 1992. "Understanding and Overcoming the NIMBY Syndrome." *Journal of the American Planning Association* 58 (Summer): 288.

Deitrick, Sabina, and Cliff Ellis. 2004. "New Urbanism in the Inner City." *Journal of the American Planning Association*. 70(4): 426.

Delogu, Orlando E. 1990. "'NIMBY' Is a National Environmental Problem." *South Dakota Law Review* 35: 1,998.

Desfor, Gene, and Roger Keil. 2004. *Nature and the City: Making Environmental Policy in Toronto and Los Angeles*. Tucson: University of Arizona Press.

Dowdell, Jennifer, Harrison Fraker, and Joan Nassauer. 2005. "Replacing a Shopping Center with an Ecological Neighborhood." *Places: Forum of Design for the Public Realm*. 17(3): 66.

Dreier, Peter. 2006. "Katrina and Power in America." *Urban Affairs Review* 41, no. 4: 528–49.

Dubin, Jon C. 1993. "From Junkyards to Gentrification: Explicating a Right to Protective Zoning in Low-Income Communities of Color." *Minnesota Law Review* 77: 739.

Dunn, James R., et al. 2006. "Housing as a Socio-Economic Determinant of Health." *Canadian Journal of Public Health* 97(3): S11.

Ellickson, Robert C., and A. Dan Tarlock. 1981. *Land-Use Controls: Cases and Materials*. Gaithersburg, Md.: Aspen Law and Business.

Ellickson, Robert C., and Vicki Been. 2005. *Land-Use Controls: Cases and Materials*. 3^d edition. New York: Aspen Publishers.

Faber, Daniel, Penn Loh, and James Jennings. 2002. "Solving Environmental Injustices in Massachusetts: Forging Greater Community Participation in the Planning Process." *Projections: The Massachusetts Institute of Technology Journal of Planning* 3, no. 2: 109.

Fagence, Michael. 1977. *Citizen Participation in Planning*. Oxford: Pergamon Press.

Fahsbender, John. 1996. "An Analytical Approach to Defining the Affected Neighborhood in the Environmental Justice Context." *New York University Environmental Law Journal* 5: 120.

Farber, Daniel A., and Jim Chen. 2006. *Disasters and the Law: Katrina and Beyond*. New York: Aspen Publishers, Inc.

Felten, Jennifer. 2006. "Brownfield Redevelopment 1995-2005: An Environmental Justice Success Story?" *Real Property, Probate and Trust Journal* 40: 679.

Folger, Robert. 1977. "Distributive and Procedural Justice: Combined Impact of "Voice" and Improvement on Experienced Inequity." *Journal of Personality and Social Psychology* 35, no. 2: 108.

Forester, John. 1989. *Planning in the Face of Power*. Berkeley: University of California Press.

-----. 2001. *The Deliberative Practitioner: Encouraging Participatory Planning Processes*. Cambridge, Mass.: MIT Press.

Foster, Sheila. 1999a. "Public Participation." In *The Law of Environmental Justice: Theories and Procedures to Address Disproportionate Risks*, edited by Michael B. Gerrard. Chicago: American Bar Association.

-----. 1999b. "Impact Assessment." In *The Law of Environmental Justice: Theories and Procedures to Address Disproportionate Risks*, edited by Michael B. Gerrard. Chicago: American Bar Association.

France, Robert L. ed. 2002. *Handbook of Water Sensitive Planning and Design*. Boca Raton, Fla.: CRC Press.

Fukuyama, Francis. 1995. *Trust: The Social Virtues and the Creation of Prosperity*. New York: The Free Press.

Galster, George, Peter Tatian, and John Accordino. 2006. "Targeting Investments for Neighborhood Revitalization." *Journal of the American Planning Association* 72(4): 457.

Garcia, Robert, and Erica Flores. 2005. "Anatomy of the Urban Parks Movement: Equal Justice, Democracy, and Livability in Los Angeles." In Robert D. Bullard (Ed.), *The Quest for Environmental Justice: Human Rights and the Politics of Pollution*. San Francisco: Sierra Club Books.

Garcia, Robert, Erica S. Flores, and Elizabeth Pine. 2002. *Dreams of Fields: Soccer, Community, and Equal Justice*. Los Angeles: The City Project, Center for Law in the Public Interest. www.cityprojectca.org/pdf/dreamsoffields.pdf

Garcia, Robert, Erica S. Flores, Katrina McIntosh, and Elizabeth Pine. 2004. *The Heritage Parkscape in the Heart of Los Angeles*. Los Angeles: The City Project, Center for Law in the Public Interest.

Garcia, Robert, and Aubrey White. 2006. *Healthy Parks, Schools, and Communities: Mapping Green Access and Equity for the Los Angeles Region*. Los Angeles: The City Project.

Garvin, Alexander. 2000. *Parks, Recreation, and Open Space: A Twenty-First Century Agenda*. Planning Advisory Service Report No. 497/498. Chicago: American Planning Association.

Gauna, Eileen. 1995. "Federal Environmental Citizen Provisions: Obstacles and Incentives on the Road to Environmental Justice." *Ecology Law Quarterly* 22: 1.

-----. 2002. "An Essay on Environmental Justice: The Past, the Present, and Back to the Future." *Natural Resources Journal* 42: 701.

-----. 2005. "Environmental Justice in a Dryland Democracy: A Comment on Water Basin Institutions." In *Wet Growth: Should Water Law Control Land-use?*, edited by Craig Anthony (Tony) Arnold. Washington, D.C.: Environmental Law Institute.

Gerrard, Michael B. 1999. *The Law of Environmental Justice: Theories and Procedures to Address Disproportionate Risks*. Chicago: American Bar Association.

-----. 2001. "Environmental Justice and Local Land-use Decisionmaking." In *Trends in Land-use Law from A to Z*, edited by Patricia E. Salkin. Chicago: American Bar Association.

Gies, Erica. 2006. *The Health Benefits of Parks: How Parks Keep Americans and Their Communities Fit and Healthy*. San Francisco: The Trust for Public Land.

Gilman, Michel Estrin. 2005. "Poverty and Communitarianism: Toward a Community-Based Welfare System." *University of Pittsburgh Law Review* 66: 721.

Goodno, James B. 2002. "Affordable Housing: Who Pays Now?" *Planning*. November: 4.

Gordon-Larsen, Penny. 2006. "Inequality in the Built Environment Underlies Key Health Disparities in Physical Activity and Obesity." *Pediatrics* (February): 417.

Gottlieb, Robert, and Andrea Misako Azuma. 2005. "Re-envisioning the Los Angeles River: An NGO and Academic Institute Influence the Policy Discourse." *Golden Gate University Law Review* 35(3): 321.

Granado, Lorraine. 1997. Executive Director, Colorado People's Environmental and Economic Network, Denver, Colorado. Telephone Interview. 21 and 22 July.

Greenberger, Scott S. 1997. "A Legacy of Zoning Bias: East Austinites Seek to Reform Land-use Rules of 1931." *Austin American-Statesman*, July 21, A1.

Haar, Charles M., and Daniel William Fessler. 1986. *The Wrong Side of the Tracks*. New York: Simon and Schuster.

Hamilton, James T. 1993. "Politics and Social Costs: Estimating the Impact of Collective Action on Hazardous Waste Facilities." *Rand Journal of Economics* 24: 101.

Hanes, Grayson P., and J. Randall Minchew. 1989. "On Vested Rights to Land-use and Development." *Washington and Lee Law Review* 46: 373.

Harnik, Peter. 2006. *The Excellent City Park System: What Makes It Great and How to Get There*. San Franciso: The Trust for Public Land.

Harwood, Stacy Anne. 2003. "Environmental Justice on the Streets: Advocacy Planning as a Tool to Contest Environmental Racism." *Journal of Planning Education and Research* 23: 24.

Haurwitz, Ralph K.M., et al. 1997. "An Industrial Chokehold: Toxic Hazards Abound in East Austin, and It's No Coincidence." *Austin American-Statesman*, July 20, A1.

Heberle, Lauren. 2006. *Connecting Smart Growth and Brownfields Redevelopment*. Louisville, Ky.: University of Louisville School of Urban and Public Affairs.

Heiman, Michael. 1990. "From 'Not in My Backyard' to 'Not in Anyone's Backyard!': Grassroots Challenge to Hazardous Waste Facility Siting." *Journal of the American Planning Association* 56 (Summer): 359.

Heynen, Nik, Harold A. Perkins, and Parama Roy. 2006. "The Political Ecology of Uneven Urban Green Space: The Impact of Political Economy on Race and Ethnicity in Producing Environmental Inequality in Milwaukee." *Urban Affairs Review* 42(1): 3.

Hirschman, Albert O. 1970. *Exit, Voice, and Loyalty: Responses to Decline in Firms, Organizations, and States*. Cambridge, Mass.: Harvard University Press.

Howland, Marie. 2003. "Private Initiative and Public Responsibility for the Redevelopment of Industrial Brownfields: Three Baltimore Case Studies." *Economic Development Quarterly* 17(4): 367.

HUD (U.S. Department of Housing and Urban Development, Office of Community Planning and Development). 2004. *Mixed-Income Housing and the HOME Program.* HUD-2003-15-CPD. Washington, D.C.: U.S. Department of Housing and Community Development.

Hutch, Daniel J. 2002. "The Rationale for Including Disadvantaged Communities in the Smart Growth Metropolitan Development Framework." *Yale Law and Policy Review* 20: 353.

Jacobs, Jane. 1961. *The Death and Life of Great American Cities.* New York: Vintage Books.

Jakowitsch, Nancy. 2002. "TEA-3 and Environmental Justice." *Second National People of Color Environmental Leadership Summit – Summit II Resource Paper Series.* Washington, DC: Summit II National Office.

Jeer, Sanjay, Megan Lewis, Stuart Meck, Jon Witten, and Michelle Zimet. 1997. *Nonpoint Source Pollution: A Handbook for Local Governments.* Planning Advisory Service Report No. 476. Chicago: American Planning Association.

Jossi, Frank. 1997. "St. Paul Overhaul." *Planning* 63(11): 18.

Kahn, Jr., Peter H. 1999. *The Human Relationship with Nature: Development and Culture.* Seattle, Wash.: The University of Washington Earth Initiative.

Kaiser, Edward J., et al. 1995. *Urban Land-use Planning.* 4th edition. Urbana, Ill.: University of Illinois Press.

Kaswan, Alice. 2003. "Distributive Justice and the Environment." *North Carolina Law Review* 81: 1031. .

Kellert, Stephen R. 2005. *Building for Life: Designing and Understanding the Human-Nature Connection.* Washington, D.C.: Island Press.

Kelly, Eric Damian. ed. 1998. *Zoning and Land-use Controls.* New York: Matthew Bender.

Kelly, Eric Damian, and Barbara Becker. 2000. *Community Planning: An Introduction to the Comprehensive Plan.* Washington, D.C.: Island Press.

Kendig, Lane. 1980. *Performance Zoning.* Chicago: American Planning Association.

Kibel, Paul Stanton. 2004. "Los Angeles' Cornfield: An Old Blueprint for New Greenspace." *Stanford Environmental Law Journal* 23(2): 275.

-----. 2006. *Access to Parkland: Environmental Justice at East Bay Parks.* San Francisco, Calif.: City Parks Project, Environmental Law and Justice Clinic, Golden Gate University School of Law.

Kob, Rose A. 2000. "Riding the Momentum of Smart Growth: The Promise of Eco-Development and Environmental Democracy." *Tulane Environmental Law Journal* 14: 139.

Krieger, James, and Donna Higgins. 2002. "Housing and Health: Time Again for Public Health Action." *American Journal of Public Health* 92: 758.

Krumholz, Norman, and John Forester. 1990. *Making Equity Planning Work: Leadership in the Public Sector.* Philadelphia: Temple University Press.

Krumholz, Norman, and Pierre Clavel. 1994. *Reinventing Cities: Equity Planners Tell Their Stories.* Philadelphia: Temple University Press.

Kuehn, Robert R. 2000. "A Taxonomy of Environmental Justice." *Environmental Law Reporter* 30: 10,681.

Kushner, James A. 1979. "Apartheid in America: An Historical and Legal Analysis of Contemporary Racial Residential Segregation in the United States." *Howard Law Journal* 22: 547.

-----. 2002-2003. "Smart Growth, New Urbanism and Diversity: Progressive Planning Movements in America and Their Impact on Poor and Minority Ethnic Populations." *UCLA Journal of Environmental Law and Policy* 21: 45.

Lambert, Thomas, and Christopher Boerner. 1997. "Environmental Inequity: Economic Causes, Economic Solutions." *Yale Journal on Regulation* 14: 195.

Lavelle, Marianne, and Marcia Coyle. 1992. "Unequal Protection: The Racial Divide in Environmental Law, A Special Investigation." *National Law Journal*, September 21, S1.

Lazarus, Richard J. 1994. "Pursuing 'Environmental Justice': The Distributional Effects of Environmental Protection." *Northwestern University Law Review* 87: 787.

Lewis, Megan. ed. 2006. *Planning and Urban Design Standards*. New York: Wiley.

Libson, Nancy. 2005-2006. "The Sad State of Affordable Housing for Older People." *Generations* 29(4): 9.

Lindblom, Charles. 1959. "The Science of 'Muddling Through.'" *Public Administration Review* 19: 79.

Lucy, William H. 1988. "APA's Ethical Principles Include Simplistic Planning Theories." *Journal of the American Planning Association* 54, no. 2: 147.

Maantay, Juliana. 2001. "Zoning, Equity, and Public Health." *American Journal of Public Health* 91, no. 7: 1033.

Maantay, Juliana. 2002. "Zoning Law, Health, and Environmental Justice: What's the Connection?" *Journal of Law, Medicine, and Ethics* 30, no. 4: 572.

Mahaffey, Kathryn R., et al. 1982. "National Estimates of Blood Lead Levels, United States, 1976–1980." *New England Journal of Medicine* 307: 573.

Mandelker, Daniel R. 2003. *Land-use Law*. 5ᵗʰ edition. New York: Matthew Bender.

Mandelker, Daniel R., and A. Dan Tarlock. 1992. "Shifting the Presumption of Constitutionality in Land-Use Law." *Urban Lawyer* 24: 1.

Mank, Bradford C. 1995. "Environmental Justice and Discriminatory Siting: Risk-Based Representation and Equitable Compensation." *Ohio State Law Journal* 56: 329.

-----. 1999. "Title VI." In *The Law of Environmental Justice: Theories and Procedures to Address Disproportionate Risks*, edited by Michael B. Gerrard. Chicago: American Bar Association.

Martz, Wendelyn A. 1995. *Neighborhood-Based Planning: Five Case Studies*. Planning Advisory Service Report No. 455. Chicago: American Planning Association.

Massey, Douglas, and Nancy A. Denton. 1993. *American Apartheid: Segregation and the Making of the Underclass*. Cambridge, Mass.: Harvard University Press.

Maxwell, Ann, and Daniel Immergluck. 1997. *Liquorlining: Liquor Store Concentration and Community Development in Lower-Income Cook County Neighborhoods*. Chicago: Woodstock Institute.

McCulloch, Heather, and Lisa Robinson. 2001. *Sharing the Wealth: Resident Ownership Mechanisms*. Oakland, Calif.: PolicyLink.

McFarlane, Audrey G. 2006. "The New Inner City: Class Transformation, Concentration of Affluence and the Obligations of the Police Power." *University of Pennsylvania Journal of Constitutional Law* 8: 1.

Milburn, Curt. 2005a. "Top Redevelopments Have Business Partners." *Minneapolis-St. Paul Business Journal*. October 14.

-----. 2005b. "The Backbone of a Community." Paper distributed at a workshop, Incorporating Environmental Justice into Land Use Planning, American Planning Association, Chicago, Ill., November 30, 2005.

-----. 2006. "A Big Honkin' Project in St. Paul." *Brownfield News*. www.brownfieldnews.com. June 6.

Mohai, Paul. 1995. "The Demographics of Dumping Revisited: Examining the Impact of Alternate Methodologies in Environmental Justice Research." *Virginia Environmental Law Journal*. 14: 615.

Mohai, Paul, and Bunyan Bryant. 1991-1992. "Race, Poverty, and the Distribution of Environmental Hazards: Reviewing the Evidence." *Race, Poverty, and the Environment* 2, Fall/Winter: 3.

-----. 1992. "Environmental Injustice: Weighing Race and Class as Factors in the Distribution of Environmental Hazards." *University of Colorado Law Review* 63: 921.

Morris, Marya. 2006a. *Integrating Planning and Public Health: Tools and Strategies to Create Healthy Places.* Planning Advisory Service Report No. 539/540. Chicago: American Planning Association.

-----. 2006a. *Planning Active Communities.* Planning Advisory Service Report No. 543/544. Chicago: American Planning Association.

Moscoso, Eunice, and Ralph K.M. Haurwitz. 1997. "PODER's Woes Bigger Than Springs, Birds." *Austin American-Statesman*, July 21, A4.

NACCHO (National Association of County and City Health Officials). 2007a. *PACE-EH Guidebook.* Available at www.naccho.org/topics/environmental/CEHA.cfm.

-----. 2007b. *PACE in Practice.* Available at www.naccho.org/topics/environmental/CEHA.cfm.

NAPA (National Academy of Public Administration). 2002. *Models for Change: Efforts by Four States to Address Environmental Justice.* Washington, D.C.

-----. 2003. *Addressing Community Concerns: How Environmental Justice Relates to Land-use Planning and Zoning.* Washington, DC: National Academy of Public Administration.

Neighborhood Environmental Conference. 2002. "Phalen Corridor Initiative—Case Study from the Working with Industries for Win-Wins." www.nextstep.state.mn.us/res_detail.cfm?id=840.

New York State Department of Environmental Conservation. 2004. *Final Report of the New York State Department of Environmental Conservation Disproportionate Adverse Environmental Impact Analysis Work Group.* Albany: New York State Department of Environmental Conservation.

Nicholls, Sarah. 2001. "Measuring the Accessibility and Equity of Public Parks: A Case Study Using GIS." *Managing Leisure* 6: 201.

Pastor, Manuel, and Rachel Rosner. 2002. "Communities Armed with Buckets Take Charge of Air Quality." In *Sustainable Solutions: Building Assets for Empowerment and Sustainable Development* by the Ford Foundation. www.fordfound.org/elibrary/documents/5008/001.cfm.

Peguero, Anthony A. 2006. "Latino Disaster Vulnerability." *Hispanic Journal of Behavioral Sciences* 28 (1): 5–22.

Peiken, Matt. 2005. "Boulevard of Believers." *Knight Ridder Tribune Business News.* October 11.

Perkins, Harold A., Nik Heynen, and Joe Wilson. 2004. "Inequitable Access to Urban Reforestation: The Impact of Urban Political Economy on Housing Tenure and Urban Forests." *Cities* 21(4): 291.

The Phalen Corridor. 2005. *The Phalen Corridor Today: A Community Report.* St. Paul, Minn.: The Phalen Corridor.

The Phalen Corridor website, www.phalencorridor.org.

Powell, Lisa M. et al. 2004. "The Relationship Between Community Physical Activity Settings and Race, Ethnicity, and Socioeconomic Status." *Evidence-Based Preventive Medicine* 1(2): 135.

Press, Eyal. 2007. "The New Suburban Poverty." *The Nation,* April 23.

Pulido, Laura. 2000. "Rethinking Environmental Racism: White Privilege and Urban Development in Southern California." *Annals of the Association of American Geographers* 90, no. 1: 12.

Quinones, Benjamin B. 1994. "Redevelopment Redefined: Revitalizing the Central City with Resident Control." *University of Michigan Journal of Law Reform* 27: 689.

Rabin, Yale. 1990. "Expulsive Zoning: The Inequitable Legacy of Euclid." In *Zoning and the American Dream,* edited by Charles M. Haar and Jerold S. Kayden. Chicago: APA Planners Press.

Randolph, John. 2004. *Environmental Land-use Planning and Management.* Washington, D.C.: Island Press.

Rechtschaffen, Clifford. 2003. "Advancing Environmental Justice Norms." *University of California at Davis Law Review* 37: 95.

Rechtschaffen, Clifford, and Eileen Gauna. 2002. *Environmental Justice: Law, Policy, and Regulation.* Durham, N.C.: Carolina Academic Press.

Retsinas, Nicolas P., and Eric S. Belsky. eds. 2002. *Low-Income Homeownership: Examining the Unexamined Goal.* Washington, D.C.: Brookings Institution Press.

Rohe, William M., and Harry L. Watson. eds. 2007. *Chasing the American Dream: New Perspectives on Affordable Homeownership.* Ithaca, N.Y.: Cornell University Press.

Reynolds, Jr., Osborne M. 1995. "'Spot Zoning'—A Spot That Could Be Removed From the Law." *Washington University Journal of Urban and Contemporary Law* 48: 117.

Rieser, Alison. 1987. "Managing the Cumulative Effects of Coastal Land Developmetn: Can Maine Law Meet the Challenge?" *Maine Law Review* 39: 321.

Rhodes, Robert M., and Cathy M. Sellers. 1991. "Vested Rights: Establishing Predictability in a Changing Regulatory System." *Stetson Law Review* 20: 475.

Rubin, Victor. 2006. "Safety, Growth, and Equity: Infrastructure Policies That Promote Opportunity and Inclusion." www.policylink.org.

Russell, Charles. 1996. "Environmental Equity: Undoing Environmental Wrongs to Low Income and Minority Neighborhoods." *Journal of Affordable Housing and Community Development Law* 5 (Winter): 147.

Salzman, James. 1997. "Valuing Ecosystem Services." *Ecology Law Quarterly* 24: 887.

Sanchez, Thomas W., and James F. Wolf. 2005. *Environmental Justice and Transportation Equity: A Review of Metropolitan Planning Organizations.* Washington, D.C.: Brookings Institution.

Sanders, Welford, and Judith Getzels. 1987. *The Planning Commission: Its Composition and Function.* Planning Advisor Service Report No. 400. Chicago: American Planning Association.

Schill, Michael, and Susan M. Wachter. 1995. "The Spatial Bias of Federal Housing Law and Policy: Concentrated Poverty in Urban America." *University of Pennsylvania Law Review* 143: 1,285.

Schwab, Jim. 1995. "Land-Use Planning and Environmental Justice." *Environment and Development.* July, 1.

Schwartz, Alex F. 2006. *Housing Policy in the United States: An Introduction.* New York: Routledge.

Selmi, Daniel P., and James A. Kushner. 2004. *Land-use Regulation: Cases and Materials.* 2d edition. New York: Aspen Publishers.

Simon, William H. 2002. "The Community Economic Development Movement." *Wisconsin Law Review* 2002: 377.

Singer, Molly, and Adam Ploetz. 2002. *Old Tools and New Measures: Local Government Coordination of Brownfields Redevelopment for Historic and Cultural Reuses.* Washington, D.C.: International City/County Management Association.

Smith, Heather Anne, and Owen J. Furuseth. 2004. "Housing, Hispanics, and Transitioning Geographies in Charlotte, North Carolina." *Southeastern Geographer* 44(2): 216.

So, Frank S., and Judith Getzels. eds. 1988. *The Practice of Local Government Planning.* 2d edition. Washington, D.C.: International City/County Management Association.

Solitare, Laura, and Michael Greenberg. 2002. "Is the U.S. Environmental Protection Agency Brownfields Assessment Pilot Program Environmentally Just?" *Environmental Health Perspectives.* 110(Supp. 2): 249.

Spyke, Nancy Perkins. 2001. "Charm in the City: Thoughts on Urban Ecosystem Management." *Journal of Land-use and Environmental Law* 16: 153.

Squires, Gregory D. 2007. "Demobilization of the Individualistic Bias: Housing Market Discrimination as a Contributor to Labor Market and Economic Inequality." *Annals* 609: 200.

Talen, Emily. 2006. "Neighborhood-Level Social Diversity." *Journal of the American Planning Association* 72(4): 431.

thomas-houston, marilyn m., and Mark Schuller. 2006. *Homing Devices: The Poor as Targets of Public Housing Policy and Practice*. Lanham, Md.: Lexington Books.

Tsoulas, Konstantinos, et al. 2007. "Promoting Ecosystem and Human Health in Urban Areas Using Green Infrastructure: A Literature Review." *Landscape and Urban Planning* 81: 167.

United Church of Christ Commission for Racial Justice. 1987. *Toxic Wastes and Race: A National Report on the Racial and Socioeconomic Characteristics of Communities with Hazardous Waste Sites*. New York: Public Data Access, Inc.

USEPA (U.S. Environmental Protection Agency). 2002. "In St. Paul, Partnerships and Developer Incentives Are Working to Restore Two Blighted Corridors." *Brownfields Success Stories*. EPA 500-F-02-166. www.epa.gov/brownfields.

-----. 2005. *National Management Measures to Control Nonpoint Source Pollution from Urban Areas*. EPA-841-B-05-004.

-----. 2006. *Protecting Water Resources with Higher-Density Development*. EPA 231-R-06-001. Washington, D.C.: USEPA.

U.S. General Accounting Office. 1983. *Siting of Hazardous Waste Landfills and Their Correlation with Racial and Economic Status of Surrounding Communities*. Washington, DC: U.S. Government Printing Office.

Vanderwarker, Amy. 2006. "Water, Environmental Justice and Land Use Planning: Richmond, California." *Progressive Planning* 169: 26.

VanScoy, Kayte. 1997. "Residents Say Recycling Plants Consitute Enviro-racism: Eastsider Decry BFI." *Austin Chronicle*, www.auschron.com/current/pols.council.html, June 2.

Weis, Natalie E. 2007. "'Left a Good Job in the City': A Gendered Analysis of Worker Mobility Strategies in Louisville Metro." Unpublished paper presented at the Urban Affairs Annual Meeting. Seattle, Washington: April 25-28.

Wolf, Michael Allan. 1996. "Fruits of the 'Impenetrable Jungle': Navigating the Boundary between Land-Use Planning and Environmental Law." *Washington University Journal of Urban and Contemporary Law* 50: 5.

Young, Kenneth H. 1996. *Anderson's American Law of Zoning*. Deerfield, IL: Clark Boardman Callaghan.

Zielenbach, Sean. 2003. "Catalyzing Community Development: Hope VI and Neighborhood Revitalization." *Journal of Affordable Housing and Community Development Law* 13 (Fall): 40.]

Zimmerman, Rae. 1994. "Issues of Classification in Environmental Equity: How We Manage Is How We Measure." *Fordham Urban Law Journal* 21: 633.

Statutes, Zoning Codes, Regulations

Austin, Texas, City of. 1997a. Ordinance 970717-F, Section 13-2-190. July 18.

-----. 1997b. *Questions and Answers: East Austin Overlay District*.

-----. Department of Environmental and Conservation Services, Planning Division. n.d. *East Austin Land-use/Zoning Report*. www.ci.austin.tx.us/landuse/ea_text.htm.

California, State of. Government Code, Sections 65300 et seq.

-----. Governor's Office of Planning and Research. 2003. *General Plan Guidelines*. Sacramento.

Los Angeles, California, City of. 1997. Conditional Use Approval for Sale of Alcoholic Beverages Specific Plan. Ordinance No. 171,681.

Orange, California, City of. 1996. Municipal Code. November.

Pittsburgh, Pennsylvania, City of. 1996. Zoning Code.

San Antonio, Texas, City of. 1997. Unified Development Code.

Santa Ana, California, City of. 1997. Municipal Code.

Washington Substitute Senate Bill 6156, Chapter 501, Law of 2007, 60th Legislature, 2007 Regular Session, Approved May 15, 2007, except for partial veto.

Wichita and Sedgwick, Kansas, County of. 1997. Unified Zoning Code. February 13.

Judicial Opinions (alphabetically by name of case)

1000 Friends of Oregon v. Board of County Commissioners, 575 P.2d 651, 656–57 (Or. Ct. App. 1978)

Amoco Oil Co. v. Village of Schaumberg, 661 N.E.2d 380, 390 (Ill. App. Ct. 1995)

Bachman v. Zoning Hearing Board, 494 A.2d 1102 (Pa. 1985)

Bankoff v. Board of Adjustment, 875 P.2d 1138, 1142–43 (Okla. 1994)

Bell v. City of Waco, 835 S.W.2d 211 (Tex. App. 1992)

Board of Adjustment for Zoning Appeals of the City and County of Denver, Findings of Fact & Conclusions of Law, Case No. 72-95 (Sept. 19, 1995)

Buchanan v. Warley, 245 U.S. 60 (1917)

Buechel v. State Department of Ecology, 884 P.2d 910 (Wash. 1994)

Chester Residents Concerned for Quality Life v. Seif, 132 F.3d 925 (3rd Cir. 1997), *vacated and remanded*, 524 U.S. 974 (1998)

Citizens of Goleta Valley v. Board of Supervisors, 52 Cal. 3d. 553 (Cal. 1990)

City of Biloxi v. Hilbert, 597 So.2d 1276 (Miss. 1992)

City of Los Angeles v. Gage, 274 P.2d 34 (Cal. Dist. Ct. App. 1954)

Cooper v. Board of County Commissioners, 614 P.2d 947 (Idaho 1980)

County Commissioners. v. Arundel Corporation, 571 A.2d 1270 (Md. 1990), vacated by *Arundel Corporation v. County Commissioners*, 594 A.2d 95 (Md. 1991)

Davis v. City of Albuquerque, 648 P.2d 777 (N.M. 1982)

DeCoals, Inc. v. Board of Zoning Appeals, 284 S.E.2d 856, 859 (W. Va. 1981)

Dolan v. City of Tigard, 512 U.S. 374, 391 (1994)

Dube v. City of Chicago, 131 N.E.2d 9, 16 (Ill. 1955)

Dugas v. Town of Conway, 480 A.2d 71 (N.H. 1984)

East-Bibb Twiggs Neighborhood Association v. Macon Bibb Planning & Zoning Commission, 888 F.2d 1573 (11th Cir.), *opinion amended and superseded on denial of reh'g*, 896 F.2d 1264 (11th Cir. 1989)

Ehrlich v. Culver City, 911 P.2d 429, 438–39 (Cal. 1996)

El Pueblo Para el Aire y Agua Limpio v. County of Kings, [1992] 22 ENVTL. L. REP. (Envtl. L. Inst.) 20357, 20358 (Cal. Super. Ct., Dec. 30, 1991)

Fawn Builders, Inc. v. Planning Board of Town of Lewisboro, 636 N.Y.S. 2d 873 (N.Y. App. 1996)

First English Evangelical Lutheran Church v. County of Los Angeles, 258 Cal. App. 3d 1353 (1989)

Free State Recycling System v. Board of County Comm'rs, 885 F. Supp. 798 (D. Md. 1994)

Golden v. City of Overland Park, 584 P.2d 130 (Kan. 1978)

Green v. County Council, 508 A.2d 882, 891 (Del. Ch. 1986), affirmed, 516 A.2d 480 (Del. 1986)

Grimpel Associates v. Cohalan, 361 N.E.2d 1022 (N.Y. 1977)

Harbison v. City of Buffalo, 152 N.E.2d 42 (N.Y. 1958)

Hartford Park Tenants Association v. Rhode Island Department of Environmental Management, CA No. 99-3748 (R.I. Super. Ct. Oct. 3, 2005)

In re Shintech Inc., Petition on Permit Nos. 2466-VO, 2467-VO, 2468-VO (EPA, Sept. 10, 1997)

In the Bd. of Adjustment for Zoning Appeals of the City & County of Denver, Findings of Fact & Conclusions as to Law, No. 72-95 (Sept. 19, 1995)

Jafay v. Board of County Commissioners, 848 P.2d 892 (Colo. 1993)

Jurgenson v. County Court for Union County, 600 P.2d 1241 (Ore. 1979)

Kimball v. Court of Common Council, 167 A.2d 706 (Conn. 1961)

La Bonta v. City of Waterville, 528 A.2d 1262, 1265 (Me. 1987)

La Salle National Bank v. City of Chicago, 125 N.E.2d 609 (Ill. 1955)

Laidlaw Envtl. Serv., Inc. v. Board of Adjustments, No. 95-CV-4631 (Colo. Dist. Ct., July 2, 1996)

Lake Lucerne Civic Ass'n v. Dolphin Stadium Corp., 801 F. Supp. 684, 688 (S.D. Fla. 1992)

Layne v. Zoning Board of Adjustment, 460 A.2d 1088 (Pa. 1983)

Livingston Rock & Gravel Company v. County of Los Angeles, 272 P.2d 4 (1954)

Loretto v. Teleprompter Manhattan CATV Corp., 458 U.S. 419, 421 (1982)

Lucas v. South Carolina Coastal Council, 505 U.S. 1003, 1004 (1992)

McGowan v. Cohalan, 361 N.E.2d 1025 (N.Y. 1977)

McQuail v. Shell Oil Co., 183 A.2d 572, 574 (Del. 1962)

Moore v. Maloney, 321 S.E.2d 335, 338 (Ga. 1984)

Moviematic Industries v. Board of County Commissioners, 349 So.2d 667 (Fla. Dist. Ct. App. 1977)

Neuberger v. City of Portland, 603 P.2d 771 (Or. 1979)

Nollan v. California Coastal Comm'n., 483 U.S. 825, 837 (1987)

Oswalt v. County of Ramsey, 371 N.W.2d 241 (Minn. Ct. App. 1985)

Parks v. Planning & Zoning Commission, 425 A.2d 100, 103 (Conn. 1979)

Penn Cent. Transp. Co. v. City of New York, 438 U.S. 104, 124 (1978)

R.I.S.E., Inc. v. Kay, 768 F. Supp. 1144, 1149 (E.D. Va. 1991)

Rockville Fuel & Feed Company v. City of Gaithersburg, 291 A.2d 672 (Md. 1972)

Seabrooke Partners v. City of Chesapeake, 393 S.E.2d 191 (Va. 1990)

Security Environmental Systems, Inc. v. South Coast Air Quality Management District, 229 Cal. App. 3d 110 (1991)

Southern Burlington County N.A.A.C.P. v. Township of Mount Laurel, 336 A.2d 713 (N.J. 1975)

Standard Oil Co. v. City of Tallahassee, 183 F.2d 410 (5th Cir. 1950)

State v. Zack, 674 P.2d 329, 332 (Ariz. Ct. App. 1983)

Tahoe Sierra Preservation Council v. Tahoe Regional Planning Agency, 535 U.S. 302 (2002)

Udell v. Haas, 235 N.E.2d 897, 905 (N.Y. 1968)

Village of Arlington Heights v. Metropolitan Housing Development Corporation, 429 U.S. 252 (1977)

Von Lusch v. Board of County Commissioners, 330 A.2d 738 (Md. Ct. Spec. App. 1975)

Wakefield v. Kraft, 96 A.2d 27 (Md. 1953)

Zylka v. City of Crystal, 167 N.W.2d 45, 49 (Minn. 1969)

<small>MAKING GREAT COMMUNITIES HAPPEN</small>

The American Planning Association provides leadership in the development of vital communities by advocating excellence in community planning, promoting education and citizen empowerment, and providing the tools and support necessary to effect positive change.

500/501. Lights, Camera, Community Video. Cabot Orton, Keith Spiegel, and Eddie Gale. April 2001. 76pp.

502. Parks and Economic Development. John L. Crompton. November 2001. 74pp.

503/504. Saving Face: How Corporate Franchise Design Can Respect Community Identity (revised edition). Ronald Lee Fleming. February 2002. 118pp.

505. Telecom Hotels: A Planners Guide. Jennifer Evans-Crowley. March 2002. 31pp.

506/507. Old Cities/Green Cities: Communities Transform Unmanaged Land. J. Blaine Bonham, Jr., Gerri Spilka, and Darl Rastorfer. March 2002. 123pp.

508. Performance Guarantees for Government Permit Granting Authorities. Wayne Feiden and Raymond Burby. July 2002. 80pp.

509. Street Vending: A Survey of Ideas and Lessons for Planners. Jennifer Ball. August 2002. 44pp.

510/511. Parking Standards. Edited by Michael Davidson and Fay Dolnick. November 2002. 181pp.

512. Smart Growth Audits. Jerry Weitz and Leora Susan Waldner. November 2002. 56pp.

513/514. Regional Approaches to Affordable Housing. Stuart Meck, Rebecca Retzlaff, and James Schwab. February 2003. 271pp.

515. Planning for Street Connectivity: Getting from Here to There. Susan Handy, Robert G. Paterson, and Kent Butler. May 2003. 95pp.

516. Jobs-Housing Balance. Jerry Weitz. November 2003. 41pp.

517. Community Indicators. Rhonda Phillips. December 2003. 46pp.

518/519. Ecological Riverfront Design. Betsy Otto, Kathleen McCormick, and Michael Leccese. March 2004. 177pp.

520. Urban Containment in the United States. Arthur C. Nelson and Casey J. Dawkins. March 2004. 130pp.

521/522. A Planners Dictionary. Edited by Michael Davidson and Fay Dolnick. April 2004. 460pp.

523/524. Crossroads, Hamlet, Village, Town (revised edition). Randall Arendt. April 2004. 142pp.

525. E-Government. Jennifer Evans–Cowley and Maria Manta Conroy. May 2004. 41pp.

526. Codifying New Urbanism. Congress for the New Urbanism. May 2004. 97pp.

527. Street Graphics and the Law. Daniel Mandelker with Andrew Bertucci and William Ewald. August 2004. 133pp.

528. Too Big, Boring, or Ugly: Planning and Design Tools to Combat Monotony, the Too-big House, and Teardowns. Lane Kendig. December 2004. 103pp.

529/530. Planning for Wildfires. James Schwab and Stuart Meck. February 2005. 126pp.

531. Planning for the Unexpected: Land-Use Development and Risk. Laurie Johnson, Laura Dwelley Samant, and Suzanne Frew. February 2005. 59pp.

532. Parking Cash Out. Donald C. Shoup. March 2005. 119pp.

533/534. Landslide Hazards and Planning. James C. Schwab, Paula L. Gori, and Sanjay Jeer, Project Editors. September 2005. 209pp.

535. The Four Supreme Court Land-Use Decisions of 2005: Separating Fact from Fiction. August 2005. 193pp.

536. Placemaking on a Budget: Improving Small Towns, Neighborhoods, and Downtowns Without Spending a Lot of Money. December 2005. 133pp.

537. Meeting the Big Box Challenge: Planning, Design, and Regulatory Strategies. Jennifer Evans–Crowley. March 2006. 69pp.

538. Project Rating/Recognition Programs for Supporting Smart Growth Forms of Development. Douglas R. Porter and Matthew R. Cuddy. May 2006. 51pp.

539/540. Integrating Planning and Public Health: Tools and Strategies To Create Healthy Places. Marya Morris, General Editor. August 2006. 144pp.

541. An Economic Development Toolbox: Strategies and Methods. Terry Moore, Stuart Meck, and James Ebenhoh. October 2006. 80pp.

542. Planning Issues for On-site and Decentralized Wastewater Treatment. Wayne M. Feiden and Eric S. Winkler. November 2006. 61pp.

543/544. Planning Active Communities. Marya Morris, General Editor. December 2006. 116pp.

545. Planned Unit Developments. Daniel R. Mandelker. March 2007. 140pp.

546/547. The Land Use/Transportation Connection. Terry Moore and Paul Thorsnes, with Bruce Appleyard. June 2007. 440pp.

548. Zoning as a Barrier to Multifamily Housing Development. Garrett Knaap, Stuart Meck, Terry Moore, and Robert Parker. July 2007. 80pp.

549/550. Fair and Healthy Land Use: Environmental Justice and Planning. Craig Anthony Arnold. October 2007. 168pp.

For price information, please go to APA's PlanningBooks.com or call 312-786-6344.
You will find a complete subject and chronological index to the PAS Report series at www.planning.org/pas.